PACEMAK

Basic Mathematics

TEACHER'S PLANNING GUIDE

GLOBE FEARON EDUCATIONAL PUBLISHER
Upper Saddle River, New Jersey
www.globefearon.com

REVIEWERS
We thank the following educators, who provided valuable comments and suggestions during the development of this book:

Rosemarie Estok, Woodbridge Public Schools, Woodbridge, New Jersey
Audris Griffith, Glen Bard West High School, Glen, Illinois
Dorie Knaub, Downey Unified School District, Downey, California
Christine Sweat, Highland Middle School, Jacksonville, Florida

Subject Area Consultant: Kay McClain, Department of Teaching and Learning, Vanderbilt University, Nashville, Tennessee
Pacemaker Curriculum Advisor: Stephen C. Larsen, formerly of The University of Texas at Austin

Supervising Editor: Stephanie Petron Cahill
Senior Editor: Phyllis Dunsay
Editors: Dena Pollak, Theresa McCarthy
Editorial Development: WordWise, Inc.
Production Editor: Laura Benford-Sullivan
Designers: Evelyn Bauer, Jennifer Visco
Editorial Assistants: Kathy Bentzen, Wanda Rockwell
Market Manager: Katie Kehoe
Research Director: Angela Darchi
Cover Design: Evelyn Bauer
Electronic Composition: Linda Bierniak, Debbie Childers, Mimi Raihl, Phyllis Rosinsky

Copyright ©2000 by Globe Fearon Inc., One Lake Street, Upper Saddle River, New Jersey, 07458. All rights reserved. No part of this book may be reproduced or transmitted in any form or by any means, electronic, photographic, mechanical, or otherwise, including photocopying, recording, or by any information storage and retrieval system, without permission in writing from the publisher.

ISBN 0-835-95746-2

Printed in the United States of America
2 3 4 5 6 7 8 9 10 03 02 01 00

GLOBE FEARON EDUCATIONAL PUBLISHER
Upper Saddle River, New Jersey
www.globefearon.com

Contents

About the Teacher's Planning Guide	vi
More About Cooperative Group Activities	viii
More About Customizing for Individual Needs	ix
More About Assessment	x
Individual Activity Rubirc	xi
Group Activity Rubric	xii

UNIT ONE: WHOLE NUMBERS — 1

Chapter 1: Understanding Whole Numbers — 2
	Opening the Chapter	4
1.1	What Is a Whole Number?	5
1.2	Odd and Even Numbers	5
1.3	Place Value to Thousands	6
1.4	Place Value to Millions	7
	Using Your Calculator: A Place-Value Game	7
1.5	Reading and Writing Whole Numbers	8
1.6	Comparing Whole Numbers	8
1.7	Ordering Whole Numbers	9
1.8	Problem Solving: Reading Tables	10
1.9	Rounding Whole Numbers	10
	Math In Your Life: Understanding Computer Memory	11
	Closing and Assessing the Chapter	12

Chapter 2: Adding Whole Numbers — 13
	Opening the Chapter	15
2.1	What Is Addition?	15
2.2	Basic Addition	16
	Using Your Calculator: Beat the Calculator	17
2.3	Column Addition	17
2.4	Adding Larger Numbers	18
2.5	Problem Solving: Clue Words for Addition	18
2.6	Adding with One Regrouping	19
2.7	Adding with More Than One Regrouping	20
2.8	Estimating Sums	20
	On-The-Job Math: Dietician	21
	Closing and Assessing the Chapter	22

Chapter 3: Subtracting Whole Numbers — 23
	Opening the Chapter	25
3.1	What Is Subtraction?	25
3.2	Basic Subtraction	26
	Using Your Calculator: Beat the Calculator	27
3.3	Subtracting Larger Numbers	27
3.4	Problem Solving: Clue Words for Subtraction	28
3.5	Subtracting with One Regrouping	29
	Math In Your Life: Monthly Expenses	29
3.6	Subtracting with More Than One Regrouping	30
3.7	Regrouping with Zeros	31
3.8	Problem Solving: Add or Subtract?	32
	Closing and Assessing the Chapter	33

Chapter 4: Multiplying Whole Numbers — 34
	Opening the Chapter	36
4.1	What Is Multiplication?	36
4.2	Basic Multiplication	37
4.3	Multiplying Larger Numbers	38
	Using Your Calculator: Checking Multiplication	39
4.4	Multiplying with One Regrouping	39
4.5	Multiplying with More Than One Regrouping	40
4.6	Problem Solving: Clue Words for Multiplication	41
4.7	Multiplying Whole Numbers by 10, 100, 1,000	42
	On-The-Job Math: Inventory Clerk	42
4.8	Multiplying by Numbers That Contain Zero	43
4.9	Problem Solving: Two-Part Problems	44
	Closing and Assessing the Chapter	45

Chapter 5: Dividing Whole Numbers — 46
	Opening the Chapter	48
5.1	What Is Division?	49
5.2	Basic Division	49
5.3	Dividing with Remainders	50
5.4	Dividing Larger Numbers	51
5.5	Checking Division	51
	Using Your Calculator: Checking Division	52
5.6	Problem Solving: Clue Words for Division	52
5.7	Dividing by Numbers with More Than One Digit	53
5.8	Zeros in the Quotient	54
5.9	Problem Solving: Choose the Operation	54
	Math In Your Life: Determining Miles Per Gallon	55
5.10	Estimating and Thinking	56
	Closing and Assessing the Chapter	58

Chapter 6: More About Numbers — 59
	Opening the Chapter	61
6.1	Divisibility Tests for 2, 5, and 10	61
6.2	Divisibility Tests for 3, 6, and 9	62
6.3	Divisibility Test for 4	63
	Using Your Calculator: Test for Divisibility	63
6.4	Factors and Greatest Common Factor	64
6.5	Multiples and Least Common Multiple	64
6.6	Prime Numbers	65
6.7	Exponents	66
6.8	Squares and Square Roots	67
	On-The-Job Math: Electric Meter Reader	68
6.9	Problem Solving: Extra Information	68
	Closing the Chapter	69
	Assessing the Chapter	70

UNIT TWO: FRACTIONS — 71

Chapter 7: Fractions and Mixed Numbers — 72
Opening the Chapter — 74
7.1 What Is a Fraction? — 75
7.2 Recognizing Equivalent Fractions — 75
7.3 Reducing Fractions to Lowest Terms — 76
7.4 Changing Fractions to Higher Terms — 77
7.5 Finding Common Denominators — 78
7.6 Comparing Fractions — 79
7.7 Ordering Fractions — 79
Using Your Calculator: Comparing Fractions — 80
7.8 Changing Fractions to Mixed Numbers — 81
7.9 Changing Mixed Numbers to Fractions — 81
Math In Your Life: Cooking — 82
7.10 Ordering Numbers You Know — 83
7.11 Problem Solving: Patterns — 83
Closing the Chapter — 84
Assessing the Chapter — 85

Chapter 8: Multiplying and Dividing Fractions — 86
Opening the Chapter — 88
8.1 Multiplying Fractions — 89
8.2 Canceling — 90
8.3 Multiplying Fractions and Whole Numbers — 90
Using Your Calculator: Multiplying Fractions and Whole Numbers — 91
8.4 Multiplying Mixed Numbers — 92
8.5 Dividing by Fractions — 93
8.6 Dividing Whole Numbers — 94
On-The-Job Math: Car Rental Agent — 94
8.7 Problem Solving: Solve a Simpler Problem — 95
8.8 Dividing Mixed Numbers — 96
8.9 Dividing Mixed Numbers by Mixed Numbers — 97
8.10 Problem Solving: Does the Answer Make Sense? — 98
Closing and Assessing the Chapter — 99

Chapter 9: Adding and Subtracting Fractions — 100
Opening the Chapter — 103
9.1 Adding and Subtracting Like Fractions — 103
9.2 Adding Like Mixed Numbers — 104
9.3 Subtracting Like Mixed Numbers — 105
9.4 Subtracting from a Whole Number — 106
9.5 Adding Unlike Fractions — 107
Math In Your Life: Pay Day — 108
9.6 Subtracting Unlike Fractions — 108
Using Your Calculator: Finding Common Denominators — 109
9.7 Adding Unlike Mixed Numbers — 110
9.8 Subtracting Unlike Mixed Numbers — 110
9.9 Problem Solving: Multi-Part Problems — 111
Closing the Chapter — 112
Assessing the Chapter — 113

UNIT THREE: OTHER TYPES OF NUMBERS — 114

Chapter 10: Decimals — 115
Opening the Chapter — 117
10.1 What Is a Decimal? — 118
10.2 Reading and Writing Decimals — 119
10.3 Comparing Decimals — 119
10.4 Ordering Decimals — 120
10.5 Adding Decimals — 121
10.6 Subtracting Decimals — 121
10.7 Multiplying Decimals — 122
10.8 Multiplying Decimals by 10, 100, 1,000 — 123
On-The-Job Math: Bank Teller — 124
10.9 Dividing Decimals by Whole Numbers — 124
10.10 Dividing Decimals by Decimals — 125
10.11 Dividing Decimals by 10, 100, 1,000 — 126
Using Your Calculator: Number Patterns — 126
10.12 Problem Solving: Multi-Part Decimal Problems — 127
10.13 Renaming Decimals as Fractions — 127
10.14 Renaming Fractions as Decimals — 128
10.15 Rounding Decimals — 128
Closing the Chapter — 129
Assessing the Chapter — 130

Chapter 11: Percents — 131
Opening the Chapter — 133
11.1 What Is a Percent? — 133
11.2 Changing Percents to Decimals — 134
11.3 Finding the Part — 135
11.4 Sales Tax — 135
11.5 Discounts — 136
11.6 Commissions — 137
Math In Your Life: Tipping — 137
11.7 Changing Decimals to Percents — 138
11.8 Changing Fractions to Percents — 139
Using Your Calculator: The Percent Key — 139
11.9 Finding the Percent — 140
11.10 Finding the Percent Increase/Decrease — 140
11.11 Finding the Whole — 141
11.12 Finding the Original Price — 142
11.13 Problem Solving: Finding the Part, Percent, or Whole — 143
Closing and Assessing the Chapter — 144

Chapter 12: Ratios and Proportions — 145
Opening the Chapter — 147
12.1 What Is a Ratio? — 147
12.2 What is a Proportion? — 148
12.3 Solving Proportions — 149
Using Your Calculator: Solving Proportions — 149
12.4 Multiple Unit Pricing — 150
12.5 Scale Drawings — 150
On-The-Job Math: Courier — 151
12.6 Problem Solving: Using Proportions — 152
Closing and Assessing the Chapter — 153

UNIT FOUR: MEASUREMENT		
AND GEOMETRY		**154**
Chapter 13: Graphs and Statistics		**155**
	Opening the Chapter	157
13.1	Pictographs	158
13.2	Single Bar Graphs	159
13.3	Double Bar Graphs	159
13.4	Single Line Graphs	160
13.5	Double Line Graphs	161
13.6	Problem Solving: Choosing a Scale	162
13.7	Circle Graphs	163
	Math In Your Life: Making a Budget	163
13.8	Mean (Average)	164
13.9	Median and Mode	165
13.10	Histograms	165
13.11	Probability	166
	Using Your Calculator: Showing Probability	167
	Closing the Chapter	168
	Assessing the Chapter	169

Chapter 14: Customary Measurement		**170**
	Opening the Chapter	171
14.1	Length	172
14.2	Weight	173
14.3	Capacity (Liquid Measure)	174
14.4	Time	175
	Using Your Calculator: How Old Are You?	175
14.5	Problem Solving: Working with Units of Measure	176
14.6	Elapsed Time	176
14.7	Temperature	177
	On-The-Job Math: Picture Framer	177
	Closing the Chapter	178
	Assessing the Chapter	179

Chapter 15: Metric Measurement		**180**
	Opening the Chapter	182
15.1	What Is the Metric System?	182
	Math In Your Life: Better Buy	183
15.2	Length	184
15.3	Mass	185
15.4	Capacity (Liquid Measure)	185
15.5	Comparing Metric and Customary Measurements	186
	Using Your Calculator: Changing Measurements	187
15.6	Problem Solving: Two-Part Problems	187
	Closing and Assessing the Chapter	189

Chapter 16: Geometry		**190**
	Opening the Chapter	192
16.1	Points and Lines	193
16.2	Measuring Angles	193
16.3	Drawing Angles	194
	On-The-Job Math: Physical Therapist	195
16.4	Angles in a Triangle	196
16.5	Polygons	197
16.6	Perimeter	197
16.7	Areas of Squares and Rectangles	198
16.8	Area of Parallelograms	199
	Using Your Calculator: Finding Perimeter and Area	199
16.9	Area of Triangles	200
16.10	Circumference of Circles	201
16.11	Area of Circles	201
16.12	Problem Solving: Subtracting to Find Area	202
16.13	Volume of Prisms	203
16.14	Volume of Cylinders	203
	Closing the Chapter	204
	Assessing the Chapter	205

UNIT FIVE: ALGEBRA		**206**
Chapter 17: Integers		**207**
	Opening the Chapter	209
17.1	What Is an Integer?	209
17.2	Adding Integers with Like Signs	210
17.3	Adding Integers with Unlike Signs	211
17.4	Subtracting Integers	211
	Using Your Calculator: Keying In Integers	212
17.5	Multiplying Integers	213
	Math In Your Life: Wind Chill Temperature	213
17.6	Dividing Integers	214
17.7	Problem Solving: Using Integers	215
	Closing and Assessing the Chapter	216

Chapter 18: Algebra		**217**
	Opening the Chapter	219
18.1	What Is an Equation?	219
18.2	Using Parentheses	220
	On-The-Job Math: Computer Programmer	221
18.3	Order of Operations	221
18.4	Solving Equations with Addition and Subtraction	222
18.5	Solving Equations with Multiplication and Division	223
18.6	Problem Solving: Using a One-Step Equation	223
18.7	Solving Equations with More Than One Operation	224
	Using Your Calculator: Checking Solutions to Equations	225
18.8	Problem Solving: Using a Two-Step Equation	225
	Closing the Chapter	226
	Assessing the Chapter	227

INDEX	**228**

About the Teacher's Planning Guide

About Pacemaker®

Globe Fearon's Pacemaker® curriculum has consistently supplied students and educators with materials and techniques that are accessible, predictable, age-appropriate, and relevant. Now in the third edition of *Basic Mathematics*, six components provide a solid, well-balanced approach to teaching math content and building math skills.

The Six Components

The **Student Edition** presents content in small manageable lessons that teach single concepts. All learning is reinforced through consistent practice, review, and application. Point-of-use strategies and answers are found in the **Teacher's Answer Edition**. More review, practice, and enrichment are provided in the **Workbook** and in the **Classroom Resource Binder**. Support for diverse classroom settings is provided in the **Teacher's Planning Guide**. Answers to all materials are found in the separate **Answer Key**. Together, these six components form a complete program.

About This Guide

Benefits of the Planning Guide

Globe market research shows that owning this **Teacher's Planning Guide** can be an important component for every mathematics professional. While the Pacemaker *Basic Mathematics Teacher's Answer Edition* contains point-of-use strategies, common errors, suggested activities and answers, this guide provides much more—and it fits any basic mathematics program! The *Teacher's Planning Guide* is an innovative, comprehensive resource that brings together a wealth of ideas, allowing you to plan and customize each lesson to meet your classroom needs.

This guide will help you tailor mathematics lessons in a number of ways. You can:

- ✓ check for prior knowledge.
- ✓ measure skill mastery.
- ✓ reinforce concepts.
- ✓ encourage cooperative learning.
- ✓ reteach strategies.
- ✓ extend and enrich.
- ✓ assess in multiple formats.

Planning a chapter is made easy with convenient tools, such as Chapter at a Glance, Learning Objectives and Skills, Prerequisite Skills, Words to Know, and a Diagnostic and Placement Guide with Resource Planner.

Addressing the **mixed abilities** of a diverse classroom is possible through a variety of activities for cooperative groups, reteaching, reinforcement, and enrichment.

Meeting the **individual needs** of your students is done via customization strategies for ESL/LEP, visual, tactile, and auditory learners.

Assessing the diverse classroom can be accomplished using multiple methods, such as standard assessment tools, as well as alternative assessment ideas for lessons, chapters, and units.

Organization of the Planning Guide

This book is presented in a predictable and easy-to-use format. It is specifically organized into the units and chapters found in the Pacemaker *Basic Mathematics* curriculum. The outline below shows the main sections of this guide. A detailed discussion of each section follows.

Unit Overview
- Special Features
- Related Materials

Planning The Chapter
- Chapter at a Glance
- Learning Objectives and Skills
- Prerequisite Skills
- Diagnostic and Placement Guide with Resource Planner

Customizing the Chapter
- Opening the Chapter
- Supporting the Lessons
- Supporting the Features
- Closing the Chapter

Assessing the Chapter
- Traditional Assessment
- Alternative Assessment

Unit Assessment
- Review
- Standardized Test Prep

Unit Overview

Special Features
The special features of each chapter in a unit are detailed in a chart on the unit overview page. They include a *Portfolio Project*, *Math In Your Life* or *On-The-Job Math*, *Using Your Calculator*, and *Problem Solving*.

Related Materials
Related Globe Fearon programs for remediation and enrichment are highlighted. This includes programs for Practice and Remediation, Test Preparation, and Real-World Connections.
For more information on Globe Fearon materials and other programs, contact:
1-800-848-9500 or www.globefearon.com.

Planning the Chapter

Chapter at a Glance
Chapter at a Glance provides a quick preview of the lessons and features in that chapter. It also provides page references to the Pacemaker *Basic Mathematics* Student Edition.

Learning Objectives and Skills
A list of learning objectives identifies the key concepts and skills for students to demonstrate after completing the chapter. These objectives are identical to those listed in the Pacemaker *Basic Mathematics* Student Edition.

Life Skills, Communication Skills, and Workplace Skills are included. These specific skills are addressed in various lessons and features of the chapter.

Prerequisite Skills
This chart provides a list of key prerequisite skills that students need to successfully complete the chapter. Each skill is cross-referenced to program resources from Pacemaker *Basic Mathematics*, which can help address that skill. You may use these resources to assess prior knowledge or to provide remediation, as needed. The prerequisite skills are noted at point of use in the chapter.

Diagnostic and Placement Guide with Resource Planner
On the left side of this helpful chart, you will find an easy analysis of assessment, based on lesson exercises from the Pacemaker *Basic Mathematics* Student Edition. Each lesson is given percent accuracy scores: 50%, 65%, 80%, and 100% based on the number of problems answered correctly.

On the right side of the chart are resources from the components of Pacemaker *Basic Mathematics*, including Student Edition Extra Practice, Workbook Exercises, Teacher's Planning Guide, and Classroom Resource Binder. These resources are referenced to each lesson being taught. Icons indicate materials appropriate for ⟲ reteaching, ↓ reinforcement, and ⟳ enrichment. While planning the chapter, use this chart for diagnosis and placement, as well as for targeting skills that students need. The example below illustrates how this chart works.

> *Example:* Joe is assigned practice for Lesson 7.3, which has 40 problems. He gets 20 correct and is therefore succeeding at the 50% accuracy level. His Individual Education Plan (IEP) objective is to achieve 65% accuracy in mathematics.
>
> You know that he needs more work in reducing fractions to lowest terms. You could use the Resource Planner to assign additional practice, such as Exercise 56 from the Workbook or Review 78 from the Classroom Resource Binder.

Customizing the Chapter

Opening the Chapter
This section provides activities, vocabulary preview, and chapter objectives information to help motivate students and prepare them for the content ahead. All of these activities complement the material in the Pacemaker *Basic Mathematics* Student Edition. To make math relevant, students can complete the Photo Activity or Chapter Project. To prepare them for the chapter, students can review Words To Know and set goals using the Learning Objectives.

Supporting the Lessons
Lessons from the *Pacemaker Basic Mathematics* Student Edition are supported with:
- Prerequisite Skills.
- Lesson Objectives.
- Words to Know.
- Cooperative Group Activities (*see p. *viii*).
- Customizing for Individual Needs: ESL and Learning Styles (*see p. *ix*).
- Mixed Abilities: Reteaching, Reinforcement, or Enrichment Activities (*see p. *ix*).
- Alternative Assessment (*see p. *x*).

For more information about this topic, see the pages indicated.

Supporting the Features

Using Your Calculator, *Math in Your Life*, and *On-The-Job Math* are features in the Pacemaker *Basic Mathematics* Student Edition. They are supported by a variety of materials, including:
- Lesson Objectives.
- Hands-on Activities.
- Writing Activities.
- Role-Playing Activities.
- Practice.
- Internet Connections.

Closing the Chapter

This section provides ideas for summarizing and closing the chapter, including reviewing vocabulary and checking if goals were met. Ideas for using Test Tips are also provided. A final group activity provides for a cooperative application of the chapter content. A scoring rubric is provided to help grade the group activity. All activities compliment the material in the Pacemaker *Basic Mathematics* Student Edition.

Assessing the Chapter

Traditional Assessment

A variety of assessment tools are provided in the form of quizzes and tests in the Student Edition and in the Classroom Resource Binder.

Alternative Assessment

Many ideas are provided in the Teacher's Planning Guide for various modes of assessment. Teachers can interview students, have students apply the math concept to a small oral or written task, or ask students to make real-life connections.
(See page x for more on assessment.)

Unit Assessment

Each unit review in Pacemaker *Basic Mathematics* provides opportunities for cumulative assessment and standardized test practice. Resources for reviewing a unit are provided in the Student Edition and in the Classroom Resource Binder. Reviews are in multiple-choice format and can be used in conjunction with a Scantron sheet for standardized-test simulation. An additional Critical Thinking section provides open-ended and written-response questions.
(See page x for more on assessment.)

More About Cooperative Group Activities

Using Materials

Manipulatives are powerful learning tools in mathematics. They can be commercially produced or hand-made. Manipulatives that are suggested for use in the cooperative group activities are readily available. You can use one or more of the following as number generators:
- Number cubes
- Spinners
- Number cards

Managing Successful Groups

Students learn best when called upon to collaborate with others in small groups. They learn to make decisions, work cooperatively, negotiate conflicts, and take risks. The teacher's role is as facilitator, creating a supportive, effective learning environment.

Teacher Tips:
- Arrange groups so that all students are visible to you.
- Read aloud or post the *Student Rules* for group work shown below.
- Explain the task to be accomplished.
- Set a time limit for task completion.
- Avoid interrupting a group that is working well.
- Accept a higher noise level in the classroom.

Student Rules:
1. You are responsible for your own behavior.
2. You are expected to participate.
3. All ideas count.
4. You must help anyone in the group who asks for assistance.
5. Only ask for teacher help if all students in the group need help.

The cooperative group works best when each student participates in the activity. You may wish to assign the following roles to individual students in the group:
- Secretary/Recorder
- Coordinator/Manager
- Encourager

More About Customizing for Individual Needs

Each lesson is supported with ideas for customizing the cooperative group activity or lesson to meet the individual needs of your students. Activities can be completed during any phase of the lesson.

ESL Notes
ESL notes provide specific activities that will help you address the needs of LEP (Limited English Proficient) students. Suggestions include making word banks, highlighting key words, using manipulatives, peer tutoring, verbalizing processes, gesturing to show meaning, labeling visuals, building on prior knowledge, and relating to everyday life. These strategies are designed to address the following barriers facing LEP students.

Vocabulary Hurdles
- A new register of mathematical words is introduced and used.
  Examples: divisor, exponent, quotient
- Everyday words have new mathematical meanings.
  Examples: mean, scale, difference
- Many words have similar meanings.
  Example: Addition — add, sum, plus, total
- Concepts are expressed using symbols that can be translated in a variety of ways.
  Example: 2^5 — "5 groups of two" or "2 to the fifth power"
- Some symbols play different roles in other countries.

Syntax Hurdles
- Comparative structures are used.
  Examples: The answer is *greater than* 5. Sue is 8 years *older than* Joe.
- Numbers are used as nouns.
  Example: Thirty is five times a number.
- Use of a preposition is significant to meaning.
  Examples: 6 divided *by* 2; 6 divided *into* 12
- There is not a one-to-one correspondence between symbols and the words they represent.
  Example: Six divided by two is not $6\overline{)2}$.
- Logical connectors may be used.
  Examples: If… then, because, such that

Problem-Solving Hurdles
- Key words must be identified and interpreted.
- Comparison words must be interpreted.
- Questions can be located at the beginning, middle, or end of a problem.
- Sometimes, information is missing or extraneous.
- Reversal errors can occur when translating word problems.
  Example: For "there are three times as many boys as girls," the expression is not $3b = g$. It is $3g = b$.
- Word problems represent hypothetical situations and can be ambiguous.

Learning Styles
Learning Styles address the needs of visual, tactile, and auditory learners. Note icons.

Visual Learner: These ideas suggest activities which enable students to see the lesson concepts and processes.

Tactile Learner: These ideas suggest activities which students perform "hands on" to manipulate materials related to the lesson concepts and processes.

Auditory Learner: These ideas suggest activities which enable students to "hear and verbalize" the lesson concepts and processes.

More About Mixed Abilities

Reteaching Activities
These activities present another way to teach the lesson concept by having students use different strategies, models, or manipulatives.

Reinforcement Activities
These activities present opportunities to strengthen the concept by having students practice the mathematical concepts they've learned while performing a repetitive task or playing a game.

Enrichment Activities
Enrichment activities extend the mathematical concept by applying newly acquired knowledge to critical thinking tasks.

More About Assessment

Using Alternative Assessment

The Teacher's Planning Guide includes an Alternative Assessment for each lesson in the Student Edition. The Alternative Assessment can be used in addition to the traditional paper and pencil assessment to provide a complete picture of student achievement. This Alternative Assessment includes a performance objective with an example and an answer or possible answer.

Alternative Assessment activities are provided in the Teacher's Planning Guide for assessing each chapter's objectives.

Making Portfolios

A student's portfolio is a collection of his or her work over a year. The portfolio provides a long-term record of the student's best efforts, progress, and achievement. Some suggested items to be included in a portfolio are:

1. Book cover with design
2. Table of contents
3. Letter explaining the contents to the viewer
4. Math autobiography
5. Paper/pencil tests with student correction of math errors
6. Performance assessments
7. Homework samples
8. Journals
9. Project results
10. Teacher/student observations

The portfolio of work should be a quality presentation. The following scoring sheet should be provided for student self-assessment:

	POINTS
Cover design with name, class, title	10
Organized/neat/easy to read	20
Table of contents	10
Cover letter	10
Math autobiography	10
Includes all required sections	10
Quality presentation/demonstration of effort	20
Above and beyond minimum	10

The following is a suggested breakdown of grades for the year that incorporates both traditional and alternative assessment:

Exams, including the final	50%
Corrections of exams	5%
Class projects	20%
Group work, class participation	10%
Portfolio	5%
Homework	10%

Using Individual and Group Activity Rubrics

The chapter projects and group activities from the Student Edition can be included in a student's portfolio. Rubrics are provided on pages *xi* and *xii* to enable you to score these efforts either individually or as a group. Each rubric is based on a 10-point scale. This makes conversion to a percentile or letter grade easy to do. For example, a score of 10 is 100% or A⁺, a 9 is 90% or A⁻, 8 is 80% or B−, and so on.

The first eight criteria on the rubric are generic. The last two criteria are left blank so that the rubric can be customized for a specific activity or task. This guide gives two suggested criteria for every Chapter Project and every Group Activity from the Pacemaker *Basic Mathematics* Student Edition.

To use either of the rubrics, photocopy the blank master once. Write the two specific criteria in the last boxes. Copy this customized version, making one for each student or group of students. When students complete an activity, evaluate how well each criterion was met. Add the points, and convert to the preferred grading system.

Another way to use the rubric is to have students grade themselves. They can exchange papers with a teammate or grade their own. Once they have seen the rubric, students may find they have a better understanding of what is required of them in an activity.

Individual Activity Rubric

Name _____ Date _____

Chapter Number _____ Activity _____

Directions
Check ✓ one box in each row to finish each sentence.
Give each check ✓ the assigned number of points.
Add the points in each column. Then, add across to find the total score.

POINTS For this activity, (name) _____	10 all of the time	9 most of the time	8 half of the time	7 less than half of the time	6 none of the time
followed directions					
asked questions when help was needed					
worked independently when required					
used appropriate resources and materials					
completed assigned tasks					
showed an understanding of the content					
presented materials without errors					
explained thinking with support					
_____					
_____					
POINTS	+	+	+	+	=

TOTAL SCORE

xi

Group Activity Rubric

Name _____ Date _____

Chapter Number _____ Activity _____

Directions
Check ✓ one box in each row to finish each sentence.
Give each check ✓ the assigned number of points.
Add the points in each column. Then, add across to find the total score.

POINTS For this activity, (name) _____	**10** all of the time	**9** most of the time	**8** half of the time	**7** less than half of the time	**6** none of the time
followed directions					
participated in group discussions					
listened carefully to others					
used appropriate resources and materials					
completed assigned tasks					
showed an understanding of the content					
presented materials without errors					
explained thinking with support					
_____ _____					
_____ _____					
POINTS	+	+	+	+	=

TOTAL SCORE

Unit Overview

Unit 1 — Whole Numbers

CHAPTER 1 Understanding Whole Numbers	PORTFOLIO PROJECT Number Journal	MATH IN YOUR LIFE Understanding Computer Memory	USING YOUR CALCULATOR A Place-Value Game	PROBLEM SOLVING Reading Tables
CHAPTER 2 Adding Whole Numbers	PORTFOLIO PROJECT Logging Minutes	ON-THE-JOB MATH Dietician	USING YOUR CALCULATOR Beat the Calculator	PROBLEM SOLVING Clue Words for Addition
CHAPTER 3 Subtracting Whole Numbers	PORTFOLIO PROJECT Years Ago Timeline	MATH IN YOUR LIFE Monthly Expenses	USING YOUR CALCULATOR Beat the Calculator	PROBLEM SOLVING Clue Words for Subtraction Add or Subtract?
CHAPTER 4 Multiplying Whole Numbers	PORTFOLIO PROJECT Nutrition Label	ON-THE-JOB MATH Inventory Clerk	USING YOUR CALCULATOR Checking Multiplication	PROBLEM SOLVING Clue Words for Multiplication Two-Part Problems
CHAPTER 5 Dividing Whole Numbers	PORTFOLIO PROJECT Life Journal	MATH IN YOUR LIFE Determining Miles per Gallon	USING YOUR CALCULATOR Checking Division	PROBLEM SOLVING Clue Words for Division Choose the Operation
CHAPTER 6 More About Numbers	PORTFOLIO PROJECT Class Team	ON-THE-JOB MATH Electric Meter Reader	USING YOUR CALCULATOR Test for Divisibility	PROBLEM SOLVING Extra Information

RELATED MATERIALS

These are some of the Globe Fearon books that can be used to enrich and extend the material in this unit.

Practice & Remediation

Pacemaker® Practical Mathematics for Consumers
Help your students master basic math concepts while developing successful daily living skills.

Test Preparation

Math for Proficiency Level A
Give your students the support they need to succeed on proficiency exams through this whole-number-based program.

Real-World Math

Math On the Job
Teach basic math skills and significance in the real world by examining how workers use real math on the job.

Planning the Chapter

Chapter 1 • Understanding Whole Numbers

Chapter at a Glance

SE page	Lesson	
2		*Chapter Opener and Project*
4	1.1	What Is a Whole Number?
5	1.2	Odd and Even Numbers
6	1.3	Place Value to Thousands
8	1.4	Place Value to Millions
9		*Using Your Calculator: A Place-Value Game*
10	1.5	Reading and Writing Whole Numbers
12	1.6	Comparing Whole Numbers
14	1.7	Ordering Whole Numbers
16	1.8	Problem Solving: Reading Tables
18	1.9	Rounding Whole Numbers
21		*Math in Your Life: Understanding Computer Memory*
22		*Chapter Review and Group Activity*

Learning Objectives
- Identify whole numbers.
- Identify odd and even numbers.
- Recognize place value.
- Read and write whole numbers.
- Compare and order whole numbers.
- Round whole numbers.
- Solve problems by reading a table.
- Apply whole numbers to computer memory.

Life Skills
- Keep a journal.
- Interpret charts and tables.
- Understand computer memory.
- Make decisions.

Communication Skills
- Read numbers in newspapers, magazines, and other print materials.
- Read and analyze advertisements.
- Write a letter to request a catalog.
- Explain how to change megabytes to bytes.
- Make charts and tables.

PREREQUISITE SKILLS
To assess mastery of prerequisite skills, use a selection of exercises from any of the program resources referenced below. The same resources can be used to provide remediation if necessary.

	Program Resources for Review		
Skills	Student Edition	Workbook	Classroom Resource Binder
Counting whole numbers			Visuals 1, 4
Recognizing place value			Visual 5

Diagnostic and Placement Guide

Chapter 1 Understanding Whole Numbers

The Percent Accuracy scores are based on the number of problems in a lesson that have been answered correctly.

Resource Planner

After diagnosing your students' needs, use the correlating Program Resources for reinforcement, reteaching, or enrichment. Additional activities for customizing lessons can be found in this guide.

KEY
Reteaching = ⤴   Reinforcement = ⬇   Enrichment = ⤵

		Percent Accuracy				Program Resources			
	Lessons	50%	65%	80%	100%	Student Edition Extra Practice	Workbook Exercises	Teacher's Planning Guide	Classroom Resource Binder
1.1	What Is a Whole Number?	1	1	1	2		1	⬇ p. 5	Visual 1 *Number Lines: Whole Numbers*
1.2	Odd and Even Numbers	1	1	1	2		2	⤴ p. 5	
1.3	Place Value to Thousands	10	13	16	20	p. 415	3	⤴ p. 6	Visual 3 *Place-Value Study Tool: Whole Numbers*
1.4	Place Value to Millions	8	10	13	16	p. 415	4	⤴ p. 7	Visual 2 *Place-Value Chart: Whole Numbers* ⬇ Review 2 *Place Value to Millions*
1.5	Reading and Writing Whole Numbers	7	9	11	14		5	⬇ p. 8	Visual 2 *Place-Value Chart: Whole Numbers* Visual 3 *Place-Value Study Tool: Whole Numbers* ⬇ Review 2 *Numbers to Millions* ⬇ Practice 5 *Reading and Writing Whole Numbers*
1.6	Comparing Whole Numbers	12	16	19	24	p. 415	6	⤴ p. 8	Visual 3 *Place-Value Study Tool: Whole Numbers* ⤴ Review 3 *Comparing Whole Numbers* ⬇ Practice 6 *Comparing and Ordering Whole Numbers*
1.7	Ordering Whole Numbers	10	13	16	20		7	⤴ p. 9	Visual 3 *Place-Value Study Tool: Whole Numbers* ⬇ Practice 6 *Comparing and Ordering Whole Numbers* ⤵ Challenge 10 *Big Cities*
1.8	Problem Solving: Reading Tables	2	3	3	4		8	⤴ p. 10	⬇ Practice 7 *Problem Solving: Reading Tables*
1.9	Rounding Whole Numbers	17	21	26	33	p. 415	9	⤴ p. 10	Visual 1 *Number Lines: Whole Numbers* ⬇ Review 4 *Rounding Whole Numbers* ⬇ Mixed Practice 8 *Reading, Ordering, and Rounding Whole Numbers* ⤵ Challenge 10 *Big Cities*

Lessons	Percent Accuracy				Program Resources			
	50%	65%	80%	100%	Student Edition Extra Practice	Workbook Exercises	Teacher's Planning Guide	Classroom Resource Binder
Math in Your Life: Understanding Computer Memory	Can be used for portfolio assessment.						p. 11	Math in Your Life 9 Understanding Computer Memory Challenge 10 Big Cities
Chapter 1 Review								
Vocabulary Review	5	6.5	8	10			p. 12	Vocabulary 1
	(writing is worth 4 points)							
Chapter Quiz	14	18	21	27			p. 12	Chapter Test A, B 11, 12

Customizing the Chapter

Opening the Chapter
Student Edition pages 2–3

Photo Activity

Discuss large numbers in the context of concerts. Ask students in what other ways large numbers may be involved, such as the number of miles a group travels from one concert to the next, or the number of seats available in a concert hall. Have students find the number of seats available for local concert halls. Students can also use an almanac to find the total number of seats for other concert halls or arenas. Ask students how many tickets would be sold if a concert were sold out.

Words to Know

Review with the students the Words to Know on page 3 of the Student Edition. Help students remember these terms by asking them to group the terms by function, such as terms that describe a number and terms that are things you can do to a number. The following words and definitions are covered in this chapter.

whole numbers 0, 1, 2, 3, 4, 5, 6, 7, and so on

number line numbers in order shown as points on a line

even numbers numbers that end in 0, 2, 4, 6, or 8

odd numbers numbers that end in 1, 3, 5, 7, or 9

digits the symbols used to write numbers: 0, 1, 2, 3, 4, 5, 6, 7, 8, and 9

rename to show a number another way; to show place value, 28 can be renamed as 2 *tens* + 8 *ones*

rounding changing a number to the nearest ten, hundred, thousand, or so on

Number Journal Project

Summary: Students keep a daily journal in which they list and describe whole numbers they see on containers and signs, in newspapers and magazines, and other places.

Materials: newspapers, magazines, containers, almost any print material, paper, pencils.

Procedure: Students can complete this project in class with all materials provided. Or, students can keep a journal throughout this chapter and cite examples found during their regular daily routine.

Assessment: Use the Individual Activity Rubric on page *xi* of this guide. Fill in the rubric with the additional information below. For this project, students should have:

- found a range of whole numbers.
- used a variety of sources.

Learning Objectives

Review the Learning Objectives on page 4 of the Student Edition before starting the chapter. Students can use the list of objectives as a learning guide. Suggest that they write the objectives in a journal or use the *Goals and Self-Check* CM6.

After each lesson, have students write an example of the skill they learned under the appropriate objective. Suggest that students use these notes as a learning guide to help them study for the chapter test.

1.1 What Is a Whole Number?
Student Edition page 4

Prerequisite Skill
- Counting whole numbers

Lesson Objective
- Identify whole numbers on a number line.

Words to Know
whole numbers, number line

Cooperative Group Activity

"Stick-It" Number Line
Materials: self-stick notes, pencils

Procedure: Prepare one self-stick note per student and one for yourself. Start with a two-digit number and count consecutively, writing one number on each note. Mix up the order of the numbers. Give each student one self-stick note. Keep the middle number for yourself. Draw a number line on the chalkboard and place your number on the number line.

- Students come to the chalkboard in small groups to put their self-stick notes in the appropriate places on the number line.
- Repeat the activity. Begin with a number other than the middle number.

Customizing the Activity for Individual Needs
ESL Help students understand the term *whole number* by asking them to imagine that they have one whole pie or one whole apple. Discuss whether they still have a whole if they cut out a piece. Have students suggest other words or terms that mean *whole*. Be sure students aren't confusing *hole* and *whole*.

Learning Styles Students can:

 draw their own reference number line with numbers 0–20 filled in.

 write numbers on index cards, then place cards in the order they would appear on a number line.

 read numbers aloud from the lowest number counting up on the number line.

Reinforcement Activity
Provide students with Visual 1 *Number Lines: Whole Numbers* or a number line having 20 points. Give them a starting number. Then have them label the points consecutively as they count.

Alternative Assessment
Students can identify a missing whole number from a spoken list.

Example: Read the following list of numbers aloud: 15, 16, 17, 19, 20. Ask, "What number is missing?"
Answer: 18

1.2 Odd and Even Numbers
Student Edition page 5

Prerequisite Skill
- Identifying and counting whole numbers

Lesson Objective
- Identify even and odd numbers.

Words to Know
even numbers, odd numbers

Cooperative Group Activity

What Am I?
Materials: number cubes

Procedure: Organize students into groups of three. Give each group two or more number cubes. Write the headings "Even" and "Odd" on the board.

- One student in the group tosses the number cubes to generate a number.
- Another student identifies the number as even or odd.
- The third student writes the number on the chalkboard under the appropriate heading.
- Students then switch roles and repeat the activity several times.
- The class as a whole then reviews the classification of the numbers on the chalkboard, making any needed corrections.

Customizing the Activity for Individual Needs
ESL To help students differentiate between the words *odd* and *even* numbers, have them create word banks on index cards with the key word on one side and the definition and examples on the other.

Learning Styles Students can:

 make a chart of even and odd numbers from 1 to 100.

 write five to ten numbers on index cards, then sort the cards into even and odd groups.

 say aloud if a number is even or odd as it is given orally.

Reteaching Activity

Give each student a different number of counters. Have them sort the counters into two equal groups. If one counter is left over, the total number of counters is odd. Otherwise, the total number is even.

Alternative Assessment

Students can decide if a spoken number is even or odd.

Example: Is 7 even or odd? Is 32 even or odd?

Answers: odd; even

1.3 Place Value to Thousands
Student Edition pages 6–7

Prerequisite Skill
- Identifying whole numbers

Lesson Objectives
- Recognize place value to thousands.
- Rename a number to show the place value of each digit.
- Write a number given the place value of each digit.
- Use models to show numbers by place value.

Words to Know
digits, rename

Cooperative Group Activity

Show Me Your Number
Materials: Visual 5 *Base 10 Blocks*, Visual 3 *Place-Value Study Tool: Whole Numbers* or a place-value chart, index cards, pencils

Procedure: Organize students into groups of four. Give each group Visual 5 *Base 10 Blocks* and four copies of Visual 3 *Place-Value Study Tool*. Then give each student one index card on which to write a 4-digit number. Collect all cards. Shuffle them and re-distribute four per group, face down. Ask each group to turn over one card.

- One student in each group reads the number aloud.
- A second student shows it with place-value blocks.
- A third student writes it in a place-value chart or in Visual 3.
- The fourth student renames the number to show the place value of each digit.
- Students repeat the activity with the remaining cards, each time reversing roles.

Customizing the Activity for Individual Needs
ESL To help students understand the word *digits*, write 4,275 on the board. Ask students to name the digit in the ones place, the tens place, the hundreds place, and the thousands place. Have students use complete sentences. "The digit in the ones place is . . ."

Learning Styles Students can:

 use a place-value chart to represent different 4-digit numbers.

 use place-value models to represent different 4-digit numbers.

 look at a 4-digit number and rename it out loud. For example, if the number is 2,318, students would rename the number as 2 thousands and 3 hundreds and 1 ten and 8 ones.

Reteaching Activity

Have students make money models having values of $100,000; $10,000; $1,000; $100, $10, and $1 or use Visual 5 *Money Models*. Students can use the money to show the value of a number.

Alternative Assessment

Students can show an understanding of place value to thousands by using place-value blocks.

Example: Use place-value blocks to represent 1,217.

Answer: Students should arrange blocks as shown.

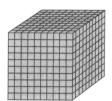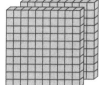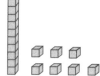

1.4 Place Value to Millions
Student Edition page 8

Prerequisite Skills
- Identifying whole numbers
- Recognizing place value to thousands

Lesson Objective
- Recognize place value to ten millions.

Cooperative Group Activity

Place-Value Flash Cards

Materials: index cards, pencils

Procedure: Organize students into pairs. Give each student 10 index cards.
- Students write a different 8-digit number on each card to make a set of place-value flash cards. Have them underline one digit in each number.
- Partners take turns showing their flash cards to each other. Each partner identifies the value of the underlined digit on each card he or she is shown.

Customizing the Activity for Individual Needs

ESL To help students understand the meaning of place value in large numbers, have them write the value of a digit as a numeral and as a word. Then have them match the numeral with its word. Each time they should say, "This is the place value of the number."

Learning Styles Students can:

 fill in Visual 2 *Place-Value Chart: Whole Numbers* to help find the values of different digits in a number.

 write each digit of a number on a separate slip of paper and mix them up. Then rearrange the slips to show that number.

 read aloud the place values in a place-value chart to help find the value of an underlined digit.

Enrichment Activity

Have students make a group scrapbook. Each student finds a number in the millions in newspapers, magazines, or a social studies book. Have the student write the number on a page in the scrapbook, tell where the number was found, and describe how the number was used. Then have the student use the number in a sentence. The student can decorate the number and write the number that comes before and after.

Alternative Assessment

Students can show an understanding of place value to millions by identifying the digit in a given place.

Example: Underline the digit in the one millions place of 42,167,809.

Answer: 42,167,809

USING YOUR CALCULATOR
A Place-Value Game
Student Edition page 9

Lesson Objectives
- Name the place value of a digit in a calculator display.
- Recognize the characteristics of a number.

Activity

Who Am I? Number Puzzles

Materials: paper, pencils, calculators

Procedure: Have students work in pairs to create and solve number puzzles.
- The first student creates a number puzzle.
- The second student guesses the number and inputs the number on the display in the calculator.

Example: I am an even number between 2,330 and 2,350. My tens digit and my ones digit are the same. Who am I?

- If the number is correct, the second student changes one digit in the number in the calculator display.
- The first student identifies the place value of the digit that was changed, then rewrites the puzzle for the new number.
- Students then reverse roles.

1.5 Reading and Writing Whole Numbers *Student Edition pages 10–11*

Prerequisite Skills
- Identifying and counting whole numbers
- Understanding place value to millions

Lesson Objectives
- Write a number using digits.
- Write a number using words.
- Communication Skill: Read whole numbers.

Cooperative Group Activity

Change That Number!

Materials: paper, index cards, pencils

Procedure: Organize students into groups of three. Give each student an index card.
- Each student writes a 3-digit number on his or her index card.
- Groups put their three cards together to form a 9-digit number. They write the number in words on a separate paper.
- Groups then rearrange the cards to form a new number and write it in words.
- Groups rearrange the cards one more time to form a third number and write it in words.

Customizing the Activity for Individual Needs

ESL Help students read and write whole numbers to millions using Visual 2 *Place-Value Chart: Whole Numbers*. Have students highlight the period names, millions and thousands, and write the number in the chart. Point out that a number within a period assumes its own name except for the ones period.

Learning Styles Students can:

 match the words in a word name with the headings on Visual 2 *Place-Value Chart*.

 make their own place-value charts and use them to help them write a number in words.

 read the word names aloud as they write them.

Reinforcement Activity

Organize students into pairs. Give each pair a number cube. One partner rolls the cube eight times, using the digits rolled to write a number. The other partner says the number and writes the number in words. Have partners switch roles and repeat. Students may continue in this manner for several rounds.

Alternative Assessment

Students can name a number orally.

Example: Read this number aloud: 98,316,411.

Answer: ninety-eight million, three hundred sixteen thousand, four hundred eleven

1.6 Comparing Whole Numbers *Student Edition pages 12–13*

Prerequisite Skills
- Identifying and counting whole numbers
- Understanding place value

Lesson Objective
- Compare whole numbers using >, <, and =.

Cooperative Group Activity

Value Sorter

Materials: Visual 12 *Hundredths Model* or large-squared grid paper, pencils, paper

Procedure: Organize students into pairs. Give each student a copy of Visual 12 or a sheet of grid paper.
- Each partner writes two different 4-digit numbers on a sheet of paper.
- Pairs exchange papers and transfer each other's numbers to grid paper. They write the digits from one number in the boxes in one row. Then they write the second number directly below the first.
- Students compare each digit from left to right. They circle the first pair of digits that are different, then circle the greater number.
- Students then write the numbers on a separate sheet of paper, placing the appropriate symbol (>, <, or =) between them.
Example: 4,321 > 3,789
- Challenge students to change one digit in either number to change the relationship.
Example: 3,321 < 3,789.

Customizing the Activity for Individual Needs

ESL Help students differentiate among the three phrases *is greater than*, *is less than*, and *is equal to* by having them match the symbols >, <, and =.

Learning Styles Students can:

 write one number below the other in Visual 12 *Hundredths Model* to compare whole numbers.

 place two numbers from 1 through ■ with the symbols >, <, and = that are written on separate index cards. Students compare numbers by arranging the appropriate cards in a formation from left to right.

 say aloud the terms "is greater than," "is less than," and "is equal to" to compare whole numbers in sentences that they read aloud.

Reteaching Activity

Have students use base ten blocks to model and compare 4-digit numbers. Students compare the number of each type of block, starting with the blocks having the largest value.

Alternative Assessment

Students can compare whole numbers by using play money to model numbers. They can model each number with play money to help them decide.

Example: Which is greater, 352 or 325?

Answer: 352

1.7 Ordering Whole Numbers
Student Edition pages 14–15

Prerequisite Skills
- Recognizing place value
- Comparing whole numbers

Lesson Objectives
- Order whole numbers.
- Communication Skill: Use and make a chart.

Cooperative Group Activity

Line Up Race

Materials: index cards, pencils

Procedure: Divide the class into teams of four or five. Distribute one index card to each student and have him or her write a number ranging from two digits to five digits on the card. Collect and shuffle the cards. Give each student one numbered index card face down. Tell students to keep their cards face down until you say, "Go."

- On *Go*, students look at their index cards. Each team finds the student with the least number on his or her card. Then team members line up in order from least to greatest behind that student. The first team to line up correctly scores 1 point.
- One student on each team retrieves everyone's cards, shuffles them, then gives the cards to another team.
- Students repeat the activity four more times. The team with the highest score at the end of five rounds wins.

Customizing the Activity for Individual Needs

ESL Pair students to build understanding of other words that mean the same as *least* and *greatest*, such as *smallest* and *largest*. Have them make a two-column chart and brainstorm words for each.

Learning Styles Students can:

 use Visual 3 *Place-Value Study Tool: Whole Numbers* or a place-value chart to help them compare and then order numbers.

 write each number on an index card or a slip of paper. Students then arrange the cards from least to greatest and from greatest to least.

 say comparisons aloud to help them order numbers. *Example:* "5 is less than 6; 6 is less than 8; 8 is greater than 5."

Reteaching Activity

Have students use play money ($1,000, $100, $10, and $1 bills) or Visual 15 *Money Models* to model and order numbers.

Alternative Assessment

Students can write numbers on individual cards and arrange the cards in order.

Example: Write the numbers 709, 1,970, and 1,907 on separate index cards. Order the numbers from least to greatest. *Answer:* 709; 1,907; 1,970

1.8 Problem Solving: Reading Tables *Student Edition pages 16–17*

Prerequisite Skills
- Identifying whole numbers
- Comparing and ordering whole numbers

Lesson Objectives
- Solve problems by reading a table.
- Make a table to solve problems.

Cooperative Group Activity

Using an Almanac
Materials: almanacs or other books of facts, index cards, pencils

Procedure: Organize students into groups of five. Give each group an almanac or a book of facts. Give each student an index card.
- Students look through the books to find a table of data.
- Each student writes a question on one side of an index card that can be answered by using the table. (He or she also writes the name of the book and the page number on the card.)
- Students write the answers to the questions on the other side of the index card.
- Students answer each other's questions by passing the cards around the group.

Customizing the Activity for Individual Needs
ESL Help students understand key words in questions about tables. Explain that *more* means that they need to compare two numbers and find the greater number. *Least* means they should find the smallest of all the numbers in the set.

Learning Styles Students can:

 use a highlighter on a copy of a table to identify those sections needed to solve a problem.

 manipulate index cards with entries from a table on them to solve the word problem.

 read aloud the questions associated with a table and explain how to use the table to answer the questions.

Enrichment Activity
Distribute magazines and newspapers to students. Ask them to look for tables. When they find a table, have them write a short paragraph describing it.

Alternative Assessment
Students can explain the information that an entry in a table provides.

Example: What does 107,200 stand for in the table?

Video Game	
Name	Points Scored
Jose	125,100
Rosemary	107,200
Beth	99,000
Mike	118,900

Answer: the number of points Rosemary scored

1.9 Rounding Whole Numbers *Student Edition pages 18–20*

Prerequisite Skills
- Identifying and counting whole numbers
- Recognizing place value
- Solving problems by reading tables

Lesson Objectives
- Round whole numbers.
- Life Skill: Interpret a table of items.

Words to Know
rounding

Cooperative Group Activity

How Close Can You Get?
Materials: Visual 1 *Number Lines: Whole Numbers* or number lines, paper, pencils

Procedure: Organize students into pairs. Distribute one copy of Visual 1 to each student.
- Each student writes a 3-digit number on paper. Pairs exchange numbers.
- Each student labels a number line with the two hundreds the given number is between and the halfway point.

Example: 321

- Students place his or her given number on the number line and determine to which hundred it is closer.
- Pairs repeat the activity by giving their partners 4-digit numbers to place on a number line and find the closer hundred.

Customizing the Activity for Individual Needs

ESL To help students say place names, have them use a place-value chart to locate the rounding place in the number. They can read and say the place names as they round.

Learning Styles Students can:

 write the given number between the rounded numbers. Then circle the nearest rounded number. For example, round 1,358 to the nearest hundred. Write 1,300; 1,358; 1,400.

 use different-colored chips on a number line to round off.

 count up or back from the original number to locate the number to which it has been rounded.

Reteaching Activity

Give 11 students name tags labeled with numbers from 50 to 60. Have them line up in order from least to greatest. Randomly pick one student and ask if the student is closer to 50 or 60. Have students explain their answer.

Alternative Assessment

Students can demonstrate their knowledge of rounding whole numbers by explaining the steps they take to round a number.

Example: Round 29,509 to the nearest thousand. Explain the steps you used.

Answer: 30,000; Student explanations should indicate knowledge that if a digit is 5 or greater, it is rounded up to the next-higher place.

MATH IN YOUR LIFE
Understanding Computer Memory *Student Edition page 21*

Lesson Objectives
- Apply whole numbers to computer memory.
- Communication Skill: Read and analyze computer advertisements.
- Communication Skill: Write a business letter to request a computer or software catalog.
- Life Skill: Make a decision about what computer and software to buy.

Activities

Look It Up
Materials: computer advertisements from newspapers, magazines, or catalogs; pencils, paper
Procedure: Have students make a table comparing the computers that are on sale, including the features of each computer, the computer memory, and the cost to add more memory.

Write a Business Letter
Materials: paper, pencils, word processor
Procedure: Have students write letters to computer and software manufacturers to request catalogs. Students can visit a computer store to look for manufacturers' addresses on software packages, or they can ask the school librarian for assistance. Students can use the catalogs to compare memory features, prices, and the amount of memory needed to install and run various software packages.

Decide What to Do
Materials: computer catalogs
Procedure: Have students use the information they have gathered to decide which computer they would like to buy and what software packages they will need. Have them consider software packages that will:
- organize their finances.
- organize their daily schedule.
- help them with their schoolwork.
- provide recreation.

Remind them that the computer they choose has to fit their needs for computer memory as well as price.

Practice

Have students complete Understanding Math in Your Life 9 *Understanding Computer Memory* from the Classroom Resource Binder.

Internet Connection

Students can use the following Web sites as starting points to shop for and find other information about computers, memory, and software, as well as jobs in the computer field.

CNET!
 http://home.cnet.com

ZD Net
 http://www.zdnet.com

Closing the Chapter
Student Edition pages 22–23

Chapter Vocabulary
Review with the students the Words to Know on page 3 of the Student Edition. Then have students quiz each other in pairs.

Have students copy and complete the Vocabulary Review questions on page 22 of the Student Edition.

For more vocabulary practice, have them complete Vocabulary 1 *Understanding Whole Numbers* from the Classroom Resource Binder.

Test Tips
Have student pairs take turns showing how to solve a problem by using one of the test tips.

Learning Objectives
Have the students review *Goals and Self-Check* CM6. They can check off the goal they have reached. Note that each section of the quiz corresponds to a Learning Objective.

Group Activity
Summary: Students locate numbers in different sections of a newspaper and compare the numbers.

Materials: newspapers, pencils, paper

Procedure: Tell each group member to review a different section of a newspaper, such as Entertainment, Home, Classifieds, and Sports. They can make a chart with four column headings: Section; Number; Less Than 1,000; Greater Than 1,000. Encourage them to use the chart to compare and analyze the numbers.

Assessment: Use the Group Activity Scoring Rubric on page xii of this guide. Fill in the rubric with the additional information below. For this activity students should have:
- compared the numbers found correctly.
- found five numbers in three different sections of the newspaper.

RELATED MATERIALS See the unit overview page for other Globe Fearon books that can be used to enrich and extend the material in this unit.

Assessing the Chapter

Traditional Assessment
Chapter Quiz
The Chapter Quiz on pages 22–23 of the Student Edition can be used as either an open- or closed-book test, or as homework. The quiz can be used to identify concepts in the chapter that students need to review and practice.

Chapter Tests
Use Chapter Test A Exercise 11 and Chapter Test B Exercise 12 *Understanding Whole Numbers* from the Classroom Resource Binder to further assess mastery of chapter concepts.

Additional Resources
Use the Resource Planner on pages 3–4 in this guide to assign additional exercises from the Classroom Resource Binder and Workbook.

Alternative Assessment
Interview
Write one 7-digit number on a piece of paper. Ask:
- What is the value of each digit?
- What is the name of this number?
- Is this number even or odd?
- What is one number with 7 digits that is greater than this number?
- Draw a line through the last three digits, making this number a 4-digit number. Round to the nearest thousand.

Journal/Portfolio
Have students find examples of large numbers in real life. These may be found in newspapers, magazines, or encyclopedia articles. Have students:
- write the numbers in words.
- tell if the numbers are odd or even.
- compare and order the numbers.
- explain how the numbers are used in real life.

Planning the Chapter

Chapter 2 • Adding Whole Numbers

Chapter at a Glance

SE page	Lesson	
24		Chapter Opener and Project
26	2.1	What Is Addition?
27	2.2	Basic Addition
28		Using Your Calculator: Beat the Calculator
29	2.3	Column Addition
30	2.4	Adding Larger Numbers
32	2.5	Problem Solving: Clue Words for Addition
34	2.6	Adding with One Regrouping
36	2.7	Adding with More Than One Regrouping
38	2.8	Estimating Sums
39		On-The-Job Math: Dietician
40		Chapter Review and Group Activity

Learning Objectives
- Add whole numbers.
- Add larger numbers.
- Add with regrouping.
- Estimate sums.
- Solve problems using addition.
- Apply addition to counting calories.

Life Skills
- Use a calculator to add.
- Find sports-related totals.
- Work backwards to solve a problem.
- Add numbers in tables and charts.
- Use the Internet and other reference materials to find information.
- Plan a vacation within a budget.

Communication Skills
- Keep a daily log.
- Write addition problems in words.
- Communicate with a patient.
- Write a letter offering suggestions.
- Explain how to find daily calorie intake.

Workplace Skills
- Estimate and calculate daily calorie intake.
- Plan and revise a menu.

PREREQUISITE SKILLS
To assess mastery of prerequisite skills, use a selection of exercises from any of the program resources referenced below. The same resources can be used to provide remediation if necessary.

Program Resources for Review

Skills	Student Edition	Workbook	Classroom Resource Binder
Reading and writing whole numbers	Lesson 1.5	Exercise 5	Practice 5
Identifying place value of whole numbers	Lessons 1.3, 1.4	Exercises 3, 4	Review 2
Rounding whole numbers	Lesson 1.9	Exercise 9	Review 4

Diagnostic and Placement Guide

**Chapter 2
Adding Whole Numbers**

The Percent Accuracy scores are based on the number of problems in a lesson that have been answered correctly.

Resource Planner

After diagnosing your students' needs, use the Correlating Program Resources for reinforcement, reteaching, or enrichment. Additional activities for customizing lessons can be found in this guide.

KEY
Reteaching = ⤺   Reinforcement = ⬇   Enrichment = ⤻

Lessons		Percent Accuracy				Program Resources			
		50%	65%	80%	100%	Student Edition · Extra Practice	Workbook Exercises	Teacher's Planning Guide	Classroom Resource Binder
2.1	What Is Addition?	8	10	12	15		10	⬇ p. 15	
2.2	Basic Addition	11	14	18	22	p. 416	11	⤺ p. 16	Visual 1 *Number Lines: Whole Numbers*
2.3	Column Addition	10	13	16	20	p. 416	12	⤺ p. 17	
2.4	Adding Larger Numbers	14	18	22	28	p. 416	13	⤺ p. 18	Visual 2 *Place-Value Chart: Whole Numbers* Visual 3 *Place-Value Study Tool: Whole Numbers*
2.5	Problem Solving: Clue Words for Addition	1	2	2	3		14	⤺ p. 18	⬇ Practice 16 *Problem Solving: Clue Words for Addition*
2.6	Adding with One Regrouping	17	21	26	33	p. 416	15	⤺ p. 19	Visual 3 *Place-Value Study Tool: Whole Numbers* ⬇ Review 14 *Adding with One Regrouping* ⬇ Practice 17 *Adding with More Than One Regrouping*
2.7	Adding with More Than One Regrouping	13	17	21	26	p. 416	16	⤺ p. 20	⬇ Review 15 *Adding with More Than One Regrouping* ⬇ Practice 17 *Adding with More Than One Regrouping* ⬇ Mixed Practice 18 *Adding Whole Numbers* ⤺ Challenge 20 *Sports on the Go*
2.8	Estimating Sums	5	7	8	10		17	⤺ p. 20	Visual 3 *Place-Value Study Tool: Whole Numbers*
	On-The-Job Math: Dietician	*Can be used for portfolio assessment.*						⤻ p. 21	⤻ On-The-Job-Math 19 *Dietician*
Chapter 2 Review									
	Vocabulary Review	6	7.5	9.5	12 *(writing is worth 4 points)*			⬇ p. 22	⬇ Vocabulary 13
	Chapter Quiz	11	14	18	22			⬇ p. 23	⬇ Chapter Test A, B 21–22

Customizing the Chapter

Opening the Chapter
Student Edition pages 24–25

Photo Activity
Procedure: Discuss numbers and addition in the context of air travel. Suggest that students take imaginary flights to two destinations. Encourage them to find the number of miles they will fly to reach each destination. They can find the information in an almanac, on a map, or in an airline flight book. Flight books are free from most major airlines. Have students round each number and then find the estimated total mileage for the two flights.

Words to Know
Review the Words to Know on page 25 of the Student Edition. Help students remember these terms by asking them to think of other words that are formed from these roots, such as addition from *add*, estimation from *estimate*, and solution from *solve*.

The following words and definitions are covered in this chapter.

add put numbers together; find the total amount

sum the amount obtained by adding; the total

plus the symbol or word that means to add

horizontal written across the page from left to right

vertical written as one thing under the other

column numbers placed one below the other

solve to find the answer to a problem

regroup to rename and then carry a tens digit to the place on the left when adding

estimate to quickly find an answer that is close to an exact answer; to make a good guess

Logging Minutes Project
Summary: Students keep a daily log of the minutes they spend doing two activities, and then add the minutes for each activity.

Materials: pencil, paper

Procedure: Students may keep a log of their activities at home or at school. They may do the calculations to find the total number of minutes in math class or as homework after completing lesson 2.7.

Assessment: Use the Individual Activity Rubric on page xi of this guide. Fill in the rubric with the additional information below. For this project, students should have:

- kept an accurate record of the number of minutes spent each day on two activities.
- calculated the daily total correctly.

Learning Objectives
Review the Learning Objectives on page 25 of the Student Edition before starting the chapter. Students can use the list as a learning guide. Suggest they write the objectives in a journal or use *Goals and Self-Check* CM6.

After each lesson, have students write an example of the skill they learned under the appropriate objective. Suggest that students use the notes as a learning guide to help them study for the chapter test.

2.1 What Is Addition?
Student Edition page 26

Prerequisite Skill
- Reading and writing whole numbers

Lesson Objectives
- Write addition problems in horizontal and vertical form using numbers.
- Communication Skill: Write addition problems in words.

Words to Know
add, sum, plus, horizontal, vertical

Cooperative Group Activity

Addition Fact Models
Materials: index cards labeled + and =, ones blocks (or counters), paper, pencils

Procedure: Organize students into groups of four. Give one card marked +, one card marked =, and several ones blocks to each group.

- One student in each group uses the ones blocks and the cards to form an addition problem. Encourage students to look for sums greater than 10. For example,

- A second student writes the addition problem in horizontal form using numbers.
- A third student writes the problem in vertical form using numbers.
- A fourth student writes the problem using words.
- Students switch roles and continue until all students have had turns creating an addition problem with the ones blocks.

Customizing the Activity for Individual Needs

ESL To increase understanding of the number words, have student write the numerals 1–29 down the side of a sheet of paper. Then have them write the English words for 1–29 next to each numeral.

Learning Styles Students can:

 draw base ten models to represent the addition facts and label each piece using symbols and words.

 manipulate ones blocks or counters to represent addition facts and record the facts using numbers horizontally and vertically.

 read addition facts aloud.

Reinforcement Activity

Provide students with flash cards showing addition facts in words. Have students write the facts in horizontal form and vertical form using numbers.

Alternative Assessment

Students can write addition problems horizontally from a spoken problem.

Example: Write this addition problem in horizontal form using numbers: "Three plus five equals eight."

Answer: 3 + 5 = 8

2.2 Basic Addition
Student Edition page 27

Prerequisite Skill
- Reading and writing addition sentences

Lesson Objective
- Use a number line to learn basic addition facts.

Cooperative Group Activity

Cards and Number Lines

Materials: index cards, Visual 1 *Number Lines: Whole Numbers* or number lines, paper, pencils

Procedure: Organize students into groups of twos or fours. Give 10 index cards and one copy of Visual 1 to each student.

- One student from each group labels the index cards 0–9 and places the cards face up on a desk.
- Another student picks two cards and shows them to the group.
- Each student in the group finds the sum of the two numbers using a number line.
- Then the group writes the numbers and the sum as an addition sentence. For example, if the numbers on the cards are 3 and 5, they write 3 + 5 = 8.
- Students take turns picking the two cards for the group.

Customizing the Activity for Individual Needs

ESL To help students understand the meaning of the word *right*, have students label the right side of the number line. As they write they say, "one is to the right of zero, two is to the right of one," and so on.

Learning Styles Students can:

 use different-colored pencils to represent the addends and the sum.

 move a finger along a number line to help them add.

 say steps aloud as they add on a number line and read their completed addition facts aloud.

Enrichment Activity

Have students decide on different ways to organize the basic addition facts. For example, they may organize them in pairs, such as 2 + 3 = 5 and 3 + 2 = 5; or group addition facts that have the same sum, such as 0 + 5 = 5, 1 + 4 = 5, 2 + 3 = 5, 3 + 2 = 5, 4 + 1 = 5, and 5 + 0 = 5.

Alternative Assessment

Students can state basic addition facts orally.

Example: What is the sum of nine and five?

Answer: 14

USING YOUR CALCULATOR
Beat the Calculator
Student Edition page 28

Lesson Objectives
- Demonstrate knowledge of basic addition facts.
- Life Skill: Use a calculator to add.

Activity
Hit the Target

Materials: calculators, paper, pencils

Procedure: Students work in pairs to choose addends for a particular sum, the target number. Each round has a different target number.

Target Numbers	Sign
Round 1: 19	
Round 2: 15	5 8
Round 3: 10	2
Round 4: 12	6 0
Round 5: 16	
Round 6: 14	7 1
Round 7: 11	9
Round 8: 17	4 3

- One player chooses two numbers from the sign whose sum is equal to the target number.
- The other player then adds the chosen numbers using the calculator. If the sum on the calculator matches the target number, the first player receives 10 points and writes the basic fact on a facts sheet.
- Roles are reversed. The second player chooses two different numbers to try to hit the same target number. Numbers may be used more than once.
- A round is completed when all the possible facts for a target number have been found.
- The player with the greatest number of points wins.

2.3 Column Addition
Student Edition page 29

Prerequisite Skills
- Reading and writing whole numbers
- Knowing basic addition facts

Lesson Objective
- Add two or more numbers in a column.

Words to Know
column

Cooperative Group Activity
Column Addition Cards

Materials: index cards, paper, pencils

Procedure: Organize students into groups of four. Give 20 index cards to each group. Instruct students to write numbers 0–9 on the first 10 index cards and numbers 0–9 on the next 10 index cards.

- One student in each group shuffles the cards and lays them face down on a desk.
- Each student in the group chooses one card and turns the card face up.
- Each student in the group writes the numbers in a column and finds the sum. Students compare their columns and their sums.
- Students repeat the activity for several rounds.

Customizing the Activity for Individual Needs

ESL To help students understand the term *column*, write the word *column* and then say it aloud. Show students how to write the numbers from a horizontal addition problem in a column. Ask students to think of other words or ways to describe the term *column*, such as *long row*.

Learning Styles Students can:

 draw circles or squares to represent the different numbers in the column. They then count the circles or squares to find the sum.

 use counters or cubes to model and solve column addition problems.

 say each step they take to solve the problem aloud. For the problem 1 + 6 + 3, they might say "one plus six equals seven; seven plus three equals ten."

Reteaching Activity

Provide each student with counters or cubes. Have students use the counters to model addition of three or more addends. Encourage students to look for different ways to find a sum. For example, the sum of 3 + 4 + 1 can be found by adding (3 + 4) + 1 or 3 + (4 + 1).

Alternative Assessment

Students can explain how to add two or more numbers.

Example: Explain how to find this sum: 5 + 3 + 1.

Possible Answer: Write the numbers in a column. Combine 5 and 3. Then add 1 to their sum.

2.4 Adding Larger Numbers
Student Edition pages 30–31

Prerequisite Skills
- Identifying place value of whole numbers
- Knowing basic addition facts

Lesson Objectives
- Add larger numbers without regrouping.
- Life Skill: Add to find total yards gained in football.

Cooperative Group Activity

Rolling Sums

Materials: blank number cubes, paper, pencils

Procedure: Organize students into groups of three. Give three number cubes to each group. Have groups write the numbers 0, 1, 2, 3, 4, and 5 on the first number cube; 80, 81, 82, 83, 90, and 91 on the second; and 600, 601, 700, 701, 800, and 801 on the third.
- Each student in the group chooses a number cube and rolls it once.
- Each group member records the numbers shown on the cubes in a column, aligning digits by place value.
- Each student finds the sum. Group members check each other's work.
- Students repeat the activity for several rounds.

Customizing the Activity for Individual Needs

ESL To help students understand the meaning of the phrase "line up the digits in each number by place value," students can use grid paper to write the numbers one under the other starting from right to left. They can name the place value of each digit of each number.

Learning Styles Students can:

 use a different color to write the digits in each addend: one color for ones, another color for tens, a third color for hundreds, and a fourth color for thousands.

 write each digit of each addend on a separate index card or slip of paper. They then align the cards in their proper places to add.

 read each digit aloud as they add each column. For 213 + 145, they may read "three plus five equals eight, one plus four equals five, two plus one equals three."

Reteaching Activity

Provide students with place-value charts or Visual 3 *Place Value Study Tool: Whole Numbers* and a spinner with four sections numbered 1 to 4. Have students spin the spinner to generate the digits for numbers with two or more digits. Have them place two numbers in the place-value columns on the chart and find the sum. As they gain confidence, students can create larger numbers.

Alternative Assessment

Students can listen to a spoken problem, write the addends in a column, and then find the sum.

Example: Find the sum of one hundred twenty-four, fifty-two, and thirteen.

Answer: 189

2.5 Problem Solving: Clue Words for Addition
Student Edition pages 32–33

Prerequisite Skills
- Identifying place value of whole numbers
- Knowing basic addition facts
- Adding larger numbers

Lesson Objectives
- Recognize clue words for addition.
- Solve word problems using addition.
- Life Skill: Work backward to solve a problem.

Words to Know
solve

Cooperative Group Activity

Words That Mean Addition

Materials: index cards, paper, pencils

Procedure: Write the following clue words on the chalkboard: *in all, together, altogether, total, both.*

Organize students into pairs. Assign each pair of students a different clue word to use.

- One student in the pair writes a word problem on the front of the index card using the assigned clue word. The other student then writes the solution on the back of the card.
- When all pairs are finished writing their problems, a volunteer collects the cards. The student places them problem-side up in a pile in the front of the room.
- Each pair picks a card from the pile and solves the problem together by identifying the clue words, then showing their addition work.
- Students check their answers by looking at the backs of the index cards.

Customizing the Activity for Individual Needs

ESL To help students understand addition clue words, have them use the words in context: *in all*, *together*, *altogether*, *total*, and *both*. Ask, "What does *in all* mean? What is another way of saying *in all*?"

Learning Styles Students can:

 circle the clue words in word problems.

 write each clue word on a slip of paper and place the papers in a pile on their desks under the addition symbol. Then, when the clue word is used in a word problem, they associate it with addition or hold up the paper.

 read each word problem to themselves, saying the clue words aloud.

Reteaching Activity

Have students underline the clue words in the problems on a worksheet or Practice 16 *Problem Solving: Clue Words for Addition*. Next have them circle the numbers to be used and solve each word problem.

Alternative Assessment

Students can make a plan to solve a word problem, then solve it.

Example: Write a plan to solve this problem, then solve it: Steven read 12 pages on Sunday, 5 pages on Monday, and 2 pages on Tuesday. How many pages did he read altogether?

Answer: The clue word *altogether* tells me to add all the pages. He read 19 pages altogether.

2.6 Adding with One Regrouping
Student Edition pages 34–35

Prerequisite Skills
- Identifying place value of whole numbers
- Adding in columns

Lesson Objectives
- Add two or three numbers with one regrouping.
- Life Skill: Add numbers in a table to calculate votes.

Words to Know
regroup

Cooperative Group Activity

Tag Team Board Addition

Materials: chalk

Procedure: Organize students into three tag teams. Write problems horizontally on the chalkboard, such as: 347 + 43, 64 + 185, and 548 + 91.

- A member from each team comes up to the chalkboard and rewrites a problem vertically.
- Tagged team members come to the board and find the sums. *Answers:* 390; 249; 639
- Tag again to come to the board and reverse the addends. Then they find the new sums. *Answers:* The sums will be the same.
- The remaining team members identify the places where regrouping is required.

Customizing the Activity for Individual Needs

ESL Help students understand the term *regroup* by writing the word, saying it aloud, and showing students how to regroup in an addition problem, such as 16 + 48.

Learning Styles Students can:

 draw squares to represent ones, tens, or hundreds digits. They then draw a large circle around ten squares to indicate regrouping where it occurs.

 use place-value blocks to add and regroup by exchanging 10 ones blocks for 1 ten block, and so on.

 explain the regrouping to a classmate while doing it. For example, students may say "nine plus three equals twelve: regroup ten ones for one ten."

Reteaching Activity

Provide students with place-value blocks. Have them use the tens and ones blocks to show 27 (2 tens, 7 ones) and 35 (3 tens, 5 ones). Ask students to combine and count the ones blocks first. If there are ten or more ones blocks, have them trade in ten ones for 1 ten block. Then ask them to combine and count to find the sum.

Alternative Assessment

Students can find the sum of two numbers with regrouping by using a place-value chart.

Example: Explain how to add 67 + 128 using a place-value chart. Then find the sum.

Answer: 195; Students should write the numbers in a place-value chart as shown. Explanations should indicate the need to regroup 15 by writing a 1 in the tens place.

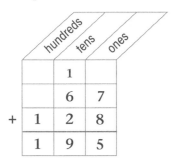

2.7 Adding with More Than One Regrouping
Student Edition pages 36–37

Prerequisite Skills
- Identifying place value of whole numbers
- Adding with one regrouping

Lesson Objectives
- Add two or three numbers with more than one regrouping.
- Life Skill: Add numbers in a chart to calculate miles traveled.

Cooperative Group Activity

Sum Announcements

Materials: paper, pencils

Procedure: Organize students into groups of three.
- One group member announces a 3-digit number greater than 600.
- Another member announces a 4-digit number greater than 5,500.
- All group members find the sum of the addends.
- Group members compare their answers.

Customizing the Activity for Individual Needs

ESL Help students understand the term *vertical*. Use gestures to explain that "in vertical form" means they should write one number on top of the other (from top to bottom) before they add.

Learning Styles Students can:
- draw triangles to represent ones digits, squares to represent tens digits, and circles to represent hundreds digits. Students then regroup by crossing out the required number of one shape and replacing the group with the new shape.
- use place-value blocks to help them add and regroup by exchanging 10 ones blocks for 1 ten block, and so on.
- explain the regrouping to the class while doing it.

Reteaching Activity

Provide students with play money ($100, $10, and $1 bills) to use to model 3-digit addition with more than one regrouping.

Alternative Assessment

Students can explain how to regroup more than once when adding.

Example: Explain the steps you would take to find why 569 + 77 = 646.

Possible Answer: Add the ones, regroup 10 ones for 1 ten, add the tens, regroup 10 tens for 1 hundred, and add the hundreds.

2.8 Estimating Sums
Student Edition page 38

Prerequisite Skill
- Rounding whole numbers

Lesson Objectives
- Estimate sums.
- Use estimation to check addition.

Words to Know
estimate

Cooperative Group Activity

Favorite Trip

Materials: almanacs or maps, paper, pencils

Procedure: Organize students into groups of three. Give an almanac or map to each group.
- Each group plans a trip to three cities. Groups use the almanac or map to determine the air or land distances between the cities.
- Students estimate the total mileage of the trip.
- Groups explain their trip to the class and show how they found the estimated mileage.

Customizing the Activity for Individual Needs

ESL Have students use their native countries for planning a trip. Allow students to find the distances in kilometers if they are more comfortable with that measurement.

Learning Styles Students can:

 draw trip stops on a map and label distances.

 trace distances on a map and use markers or pins to locate each city.

 read the city names and distances for the class.

Enrichment Activity

Have students choose a pair of items from newspaper and magazine advertisements. If the prices include cents, have them cross out the numbers following the decimal point. Have them round the prices to a specific place. Then estimate the amount of money they would need to buy those items.

Alternative Assessment

Students can estimate sums using a place-value chart.

Example: Explain why the sum shown below is incorrect. Use estimation to help.

thousands	hundreds	tens	ones
	5	7	1
	6	3	8
+	1	1	3
1	1	2	2

Answer: By rounding addends to the nearest hundred, the estimated sum is 1,300. The sum shown does not make sense. The correct sum is 1,322.

ON-THE-JOB MATH
Dietician Student Edition page 39

Lesson Objectives
- **Workplace Skills:** Use addition to calculate daily calorie intake. Plan and revise a menu.
- **Communication Skills:** Explain dietary choices to a patient. Write a letter.
- **Life Skill:** Use the Internet to find career information.

Activities

Look It Up

Materials: reference books on calories, pencils, paper

Procedure: Bring in books or other references that list the calories in various foods and the recommended daily number of calories for age and gender.
- Have students write down the foods they ate for breakfast, lunch, and dinner yesterday. Ask them to look up the number of calories for each serving of food.
- Then find the total number of calories they ate.
- Have students compare the number of calories they ate with the recommended number of calories for their ages and genders.

Write a Business Letter

Materials: paper and pens, or computer word processor

Procedure: Have students create a lunch menu of about 700 calories.
- Challenge them to make the menu nutritious by including carbohydrates, proteins, vegetables, and fruits.
- Have students write a letter to the manager of the school cafeteria requesting that their menus be considered for future lunches.

Act It Out

Materials: pencils, paper

Procedure: Give students the following scenario:
- Mr. Gomez has a heart problem. His doctor says that he must carefully choose the foods he eats. He needs a diet lower in fat and higher in complex carbohydrates, such as fruits and vegetables.
- Students role-play explaining to Mr. Gomez the changes he needs to make in his eating habits.
- Students prepare sample menus for Mr. Gomez to use for one day. The total calories for the entire day should be no more than 2,500 calories.

(Continued on p. 22)

(Continued from p. 21)

Practice
Have students complete On-The-Job Math 19 *Dietician* in the Classroom Resource Binder.

Internet Connection
The following Web sites can be used to find out about the requirements for becoming a dietician.
American Dietetic Association
   http://www.eatright.org/careers.html
Southern California Foods, Nutrition, and Dietetics Consortium
   http://www.occ.cccd.edu/organizations/scfndc/index.html

Closing the Chapter
Student Edition pages 40–41

Chapter Vocabulary
Review with the students the Words to Know on page 25 of the Student Edition. Then have students quiz each other in pairs.

Have students copy and complete the Vocabulary Review questions on page 40 of the Student Edition.

For more vocabulary practice, have them complete Vocabulary 13 *Adding Whole Numbers* from the Classroom Resource Binder.

Test Tips
In small groups, students can explain to each other how to solve a problem by using each test tip.

Learning Objectives
Have the students review *Goals and Self-Check* CM6. They can check off the goal they have reached. Note that each section of the quiz corresponds to a Learning Objective.

Group Activity
Summary: Students look through travel brochures to plan a vacation within a given budget.

Materials: brochures, newspapers

Procedure: Encourage students to brainstorm ideas of places they'd like to visit. Suggest that each student finds costs for a different category (travel, lodging, or spending money) and then check each other's work.

Assessment: Use the Group Activity Scoring Rubric on page *xii* of this guide. Fill in the rubric with the additional information below. For this activity student should have:
• stayed within all three set budget amounts.
• made plans that include all three categories.

RELATED MATERIALS See the unit overview page for other Globe Fearon books that can be used to enrich and extend the material in this unit.

Assessing the Chapter

Traditional Assessment
Chapter Quiz
The Chapter Quiz on pages 40–41 of the Student Edition can be used as either an open- or closed-book test, or as homework. The quiz can be used to identify concepts in the chapter that students need to review and practice.

Chapter Tests
Use Chapter Test A Exercise 21 and Chapter Test B Exercise 22 *Adding Whole Numbers* from the Classroom Resource Binder to further assess mastery of chapter concepts. These tests are in a multiple-choice format.

Additional Resources
Use the Resource Planner on page 14 in this guide to assign additional exercises from the Classroom Resource Binder and Workbook.

Alternative Assessment
Interview
Write 2,942 + 3,358 as a horizontal problem on paper. Ask:
• How do you round the numbers to the nearest hundred?
• What is the estimated sum?
• Look at the original numbers. Which digits will need to be regrouped when added?
• What is the exact sum?

Application
• Students write and illustrate their own word problems on poster board. Problems should require use of addition for the solution. Encourage students to use a clue word, such as *total, in all,* or *together,* in the problem.
• Students write the solution for the problem on the back of the poster.
• Posters can be displayed around the classroom.

Planning the Chapter

Chapter 3 • Subtracting Whole Numbers

Chapter at a Glance

SE page	Lesson	
42		Chapter Opener and Project
44	3.1	What Is Subtraction?
45	3.2	Basic Subtraction
46		Using Your Calculator: Beat the Calculator
47	3.3	Subtracting Larger Numbers
48	3.4	Problem Solving: Clue Words for Subtraction
50	3.5	Subtracting with One Regrouping
53		Math in Your Life: Monthly Expenses
54	3.6	Subtracting with More Than One Regrouping
56	3.7	Regrouping with Zeros
58	3.8	Problem Solving: Add or Subtract?
60		Chapter Review and Group Activity

Learning Objectives
- Subtract whole numbers.
- Subtract larger numbers.
- Subtract with regrouping.
- Subtract from zeros.
- Solve word problems using subtraction.
- Apply subtraction to monthly expenses.

Life Skills
- Use a calculator to subtract.
- Work backward to solve a problem.
- Subtract numbers in tables and charts.
- Calculate monthly expenses.
- Find passbook balances.
- Draw a diagram.

Communication Skills
- Create a timeline.
- Write subtraction problems in words.
- Explain how to estimate monthly expenses.
- Ask for information over the phone.

PREREQUISITE SKILLS
To assess mastery of prerequisite skills, use a selection of exercises from any of the program resources referenced below. The same resources can be used to provide remediation if necessary.

	Program Resources for Review		
Skills	Student Edition	Workbook	Classroom Resource Binder
Understanding place value to millions	Lesson 1.4	Exercise 4	Review 2
Comparing whole numbers	Lesson 1.6	Exercise 6	Review 3 Practice 6
Performing basic addition	Lesson 2.2	Exercise 11	
Adding larger numbers	Lesson 2.4	Exercise 13	
Adding with one regrouping	Lesson 2.6	Exercise 15	Review 14
Adding with more than one regrouping	Lesson 2.7	Exercise 16	Review 15 Practice 17 Mixed Practice 18

Diagnostic and Placement Guide

**Chapter 3
Subtracting
Whole Numbers**

The Percent Accuracy scores are based on the number of problems in a lesson that have been answered correctly.

Resource Planner

After diagnosing your students' needs, use the Correlating Program Resources for reinforcement, reteaching, or enrichment. Additional activities for customizing lessons can be found in this guide.

KEY
Reteaching = ⤺   Reinforcement = ⬇   Enrichment = ⤻

Lessons		Percent Accuracy				Program Resources			
		50%	65%	80%	100%	⬇ Student Edition • Extra Practice	⬇ Workbook Exercises	Teacher's Planning Guide	Classroom Resource Binder
3.1	What Is Subtraction?	8	10	12	15		18	⤺ p. 25	
3.2	Basic Subtraction	11	14	18	22	p. 417	19	⤺ p. 26	
3.3	Subtracting Larger Numbers	11	14	17	21	p. 417	20	⤺ p. 27	
3.4	Problem Solving: Clue Words for Subtraction	1	2	2	3		21	⬇ p. 28	⬇ Practice 26 *Clue Words for Subtraction*
3.5	Subtracting with One Regrouping	19	25	30	38	p. 417	22	⬇ p. 29	⬇ Review 25 *Subtracting with One Regrouping* ⬇ Practice 27 *Subtracting with One Regrouping*
	Math in Your Life: Monthly Expenses	*Can be used for portfolio assessment.*						⤻ p. 29	⤻ Math in Your Life 31 *Monthly Expenses*
3.6	Subtracting with More Than One Regrouping	13	16	20	25	p. 417	23	⤻ p. 30	⬇ Review 25 *Subtracting with More Than One Regrouping* ⬇ Practice 28 *Subtracting with More Than One Regrouping*
3.7	Regrouping with Zeros	17	22	27	34	p. 417	24	⤺ p. 31	⬇ Review 25 *Subtracting with More Than One Regrouping* ⬇ Practice 28 *Subtracting with More Than One Regrouping* ⤻ Challenge 32 *Keeping Count*
3.8	Problem Solving: Add or Subtract?	1	2	2	3		25	⬇ p. 32	⬇ Practice 29 *Problem Solving: Add or Subtract?* ⬇ Mixed Practice 30 *Adding and Subtracting with and without Regrouping*
Chapter 3 Review									
	Vocabulary Review	4	5.5	6.5	8			⬇ p. 32	⬇ Vocabulary 23
		(writing is worth 4 points)							
	Chapter Quiz	15	20	24	30			⬇ p. 33	⬇ Chapter Test A, B 33, 34

Customizing the Chapter

Opening the Chapter
Student Edition pages 42–43

Photo Activity
Procedure: Bring in, or have students bring to class, published sales reports of albums or music videos of popular groups and individuals. (For example, *The World Almanac and Book of Facts* lists numbers of singles, albums, and music videos that have been sold.) Suggest that students make a class chart, ordering each category from least number of sales to greatest number. Have each student select his or her favorite from the current list and determine how many more must be sold to reach the top seller.

Words to Know
Review the Words to Know on page 43 of the Student Edition. To help students remember the meaning of *minus*, point out the similarity of *min-* in *minus* with *min-* in *minimize* (to make smaller) and *minimal* (smallest).

The following words and definitions are covered in this chapter.

subtract to take away one number from another; to find the amount that remains

difference the amount obtained by subtracting; the amount by which one number is larger or smaller than another

minus the symbol or word that means "to subtract"

regroup to rename and then carry a tens digit to the place on the right when subtracting

Years Ago Timeline Project
Summary: Students create a timeline of public and personal events, then subtract to find how many years ago each event occurred.

Materials: research and reference books, magazines, adding machine tape or paper, pencils, index cards

Procedure: Suggest that students first write each event and year on index cards, then organize the cards according to year. Students can write events and dates on adding machine tape. Encourage students to list such events as year first employed in an after-school job. Students can make timelines in class with materials provided, then subtract for homework. Students can also complete the entire activity at home.

Assessment: Use the Individual Activity Rubric on page *xi* of this guide. Fill in the rubric with the additional information below. For this project, students should have:
- placed important events on the timeline correctly.
- subtracted correctly to find how many years have passed since each event.

Learning Objectives
Review the Learning Objectives on page 43 of the Student Edition before starting the chapter. Students can use the list as a learning guide. Suggest they write the objectives in a journal or use *Goals and Self-Check* CM6.

After each lesson, have students write an example of the skill they learned under the appropriate objective. Suggest that students use these notes as a learning guide to help them study for the chapter test.

3.1 What Is Subtraction?
Student Edition page 44

Prerequisite Skill
- Reading and writing whole numbers

Lesson Objectives
- Write subtraction problems in horizontal and vertical form using numbers.
- Communication Skill: Write subtraction problems in words.

Words to Know
subtract, minus, difference

Cooperative Activity

What's the Difference?
Materials: unit cubes (18 per group), index cards, pencils

Procedure: Organize the class into groups of four. Provide each group with unit cubes and index cards. Each group will create subtraction flash cards.
- One student takes a handful of unit cubes, places them in the center of the group, and counts them. The student then writes the number on an index card and passes the card to the next student.
- The next student writes a minus sign on the card and without looking takes some cubes from the center. The student counts the number of cubes that were taken away and writes the number to the right of the minus sign.

- The card passes to the next student.
- This student writes an equal sign on the card, counts the number of cubes that are left, and writes this number to the right of the equal sign. This is the difference.
- The fourth student writes the number sentence in words under the symbols.

Encourage groups to create as many subtraction cards as they can. Collect the cards and have students come together as a class. Pick a card, read the subtraction sentence, and have volunteers write the sentence in vertical form on the board.

Customizing the Activity for Individual Needs

ESL Help students differentiate between subtraction terms and addition terms. Have partners compare the words *subtract, difference,* and *minus* with the addition words *add, plus,* and *sum*. Model the words in context.

Example: "When I add, I find the *sum*." "When I subtract, I find the *difference*."

Learning Styles Students can:

 draw objects to represent subtraction problems.

 manipulate counters to represent subtraction problems.

 read subtraction problems aloud as they write the numbers.

Reteaching Activity

Have students use counters to model subtraction facts. They can create their own set of subtraction flash cards by writing the subtraction phrase on one side of a card and the difference on the other side. Encourage them to work in an orderly way to find all the subtraction facts.

Alternative Assessment

Students can write subtraction problems horizontally or vertically from a spoken problem.

Example: Write this subtraction problem using numbers: "Thirteen minus seven equals six."

Answer: 13 − 7 = 6 or $\begin{array}{r}13\\-7\\\hline 6\end{array}$

3.2 Basic Subtraction
Student Edition page 45

Prerequisite Skill
- Reading and writing subtraction sentences

Lesson Objective
- Use a number line to learn basic subtraction facts.

Cooperative Group Activity

Subtraction Spin

Materials: Visual 1 *Number Lines: Whole Numbers* or number lines, spinners with six sections, spinners with four sections, paper, pencils

Procedure: Organize the class into groups of four. Give each group a copy of Visual 1, each of the two spinners, paper, and pencils.

- One group member numbers the number line from 0 to 20 and makes two spinners. For the first spinner, the student writes one number from 11 to 16 in each of the six sections. For the second spinner, the student writes the numbers 6 to 9 in each of the four sections.
- A second student spins the first spinner and then the second spinner.
- A third student uses the number line to find the difference between the two numbers.
- A fourth student writes a subtraction sentence to record the result.
- A volunteer from each group writes the subtraction sentences on the chalkboard. Groups then compare results.

Customizing the Activity for Individual Needs

ESL Pair a second-language learner with a native English speaker to be sure that each understands how to subtract using a number line. Have one student demonstrate the procedure while the other writes the problem. Then have students switch roles.

Learning Styles Students can:

 circle the starting number and put a box around the difference before writing subtraction facts.

 move a finger along a number line to help them subtract. For example, they may put a finger on 15 and then move the finger 7 spaces to the left to end up on 8.

say steps aloud as they subtract on the number line and read their completed subtraction facts aloud.

Alternative Assessment

Students can state basic subtraction facts orally.

Example: What is twelve minus seven?

Answer: 5

Enrichment Activity

Make two sets of number cards: one with the numbers 11–18; the other with numbers less than 10. Draw a card from each set. Ask students to find the difference and write a subtraction sentence. Then ask them to write three or more subtraction sentences having the same difference. Repeat for other number combinations.

USING YOUR CALCULATOR
Beat the Calculator
Student Edition page 46

Lesson Objectives
- Demonstrate knowledge of basic subtraction facts.
- Life Skill: Use a calculator to subtract.

Activities

Hit the Target

Materials: calculators

Procedure: Students work in pairs to choose numbers for a particular difference, the target number. Each round has a different target number.

Target Numbers	Sign
Round 1: 9	
Round 2: 6	
Round 3: 4	
Round 4: 7	
Round 5: 5	
Round 6: 8	
Round 7: 3	
Round 8: 2	

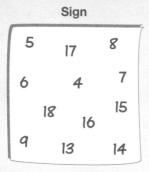

- One player chooses two numbers from the sign whose difference is equal to the target number.
- The other player then subtracts the chosen numbers using the calculator. If the difference on the calculator matches the target number, the first player receives 10 points and writes the basic fact on a facts sheet.
- Roles are then reversed. The second player chooses two different numbers to try to hit the same target number. Numbers may be used more than once.
- A round is completed when all the possible facts for a target number have been found.
- The player with the highest points wins.

Making a Difference

Materials: calculators

Procedure: One player enters a number from 6 to 14 into the calculator. The other player must then subtract a number to get a difference of exactly 5. Players are awarded one point for each correct subtraction. The numbers may be varied to obtain other goal numbers, such as using the starting numbers 10 to 18 and aiming for a difference of 9.

3.3 Subtracting Larger Numbers
Student Edition page 47

Prerequisite Skills
- Understanding place value to millions
- Knowing basic subtraction facts

Lesson Objective
- Subtract larger numbers without regrouping.

Cooperative Group Activity

Number Jumble

Materials: number cards from 1–9, paper, pencils

Procedure: Organize students into groups of three or four. Give each group a set of number cards from 1–9.

- Members take turns drawing two cards to form a 2-digit number. The group works together to subtract the 2-digit number from 99. Repeat the activity until students are comfortable subtracting 2-digit numbers from 99.
- Members take turns drawing three cards to form a 3-digit number. They then subtract from 999.
- To conclude the activity, members take turns drawing four cards to form a 4-digit number to subtract from 9,999.

Customizing the Activity for Individual Needs

ESL Pair students to review "place value." Have one student write a 2-, 3-, or 4-digit number and underline one digit. Have the other student name the place value of that digit. Students switch roles.

Learning Styles Students can:

 draw vertical lines as clues for digit placement.

 model subtraction problems with number cards, aligning the cards in their proper places to subtract.

 say each digit aloud as they subtract in each column.

Reteaching Activity

Provide students with Visual 3 *Place-Value Study Tool: Whole Numbers* or lined paper turned sideways. Write several 2- and 3-digit subtraction problems on the board in horizontal form. The subtraction problems should not require regrouping. Have students write the numbers in appropriate columns and find the differences.

Alternative Assessment

Students can identify which of several subtraction problems has an incorrect answer.

Example: Which problem has an incorrect answer?

(a) $\begin{array}{r} 857 \\ -241 \\ \hline 616 \end{array}$ (b) $\begin{array}{r} 948 \\ -635 \\ \hline 323 \end{array}$ (c) $\begin{array}{r} 766 \\ -552 \\ \hline 214 \end{array}$

Answer: **(b)**

3.4 Problem Solving: Clue Words for Subtraction
Student Edition pages 48–49

Prerequisite Skills
- Knowing basic subtraction facts
- Subtracting larger numbers

Lesson Objectives
- Recognize clue words for subtraction.
- Solve word problems using subtraction.
- Life Skill: Use addition to check subtraction.
- Life Skill: Work backward to solve a problem.

Cooperative Group Activity

Do-It-Yourself Problems
Materials: index cards, paper, pencils

Procedure: Organize students into pairs. Write the following clue words on the chalkboard: *how much more, how much less, how many more, left, remain, difference.* Assign each pair of students a different clue word to use.

- One member of the pair writes a word problem on the front of the index card using the assigned clue word. Then the other student writes the solution on the back of the card.
- When all pairs are finished writing their problems, a volunteer collects the cards and places them problem-side up in a pile in the front of the room.
- Pairs pick a card from the pile without looking. They solve the problem together by identifying the clue words, then showing their subtraction work.
- Students check their answers by looking at the backs of the index cards.

Customizing the Activity for Individual Needs

ESL Help develop understanding of subtraction clue words by pairing students in reading word problems and highlighting subtraction clue words.

Learning Styles Students can:

 underline the clue words in the problems.

 write each clue word on a slip of paper and place them in a pile on their desks. They then take the correct slip of paper out of the pile when the clue word is used in a problem.

 say aloud a subtraction sentence as a partner writes it. They then reverse roles.

Enrichment Activity

Give students a 2-digit number greater than 80, such as 87. Have students write a pattern of smaller numbers in ascending order, such as 22, 33, 44, 55, . . . , or 10, 20, 30, ... Students can then subtract each number in the pattern from the 2-digit starting number and record their answers. Ask students to describe any pattern they notice in the differences.

Answer: The differences form a sequence in descending order.

Alternative Assessment

Students can make a plan to solve a word problem. Then they solve the problem.

Example: Write a plan to solve this problem, then solve it: Van saw 12 eagles on Tuesday, but 4 flew away. How many were left?

Answer: The clue word *left* tells me to subtract. There were 8 eagles left.

3.5 Subtracting with One Regrouping
Student Edition pages 50–52

Prerequisite Skills
- Identifying place value of whole numbers
- Subtracting larger numbers

Lesson Objectives
- Subtract numbers with one regrouping.
- Life Skill: Subtract to interpret survey results in a table.

Words to Know
regroup

Cooperative Group Activity

Trading Dollars
Materials: play money ($10s and $1s), blank number cubes, paper, pencils

Procedure: Organize students into small groups. Give each group fifteen $10 bills, five $1 bills, and two blank number cubes. Have groups write the digits 0–5 on one number cube and 5–9 on the other. Have available extra $1 bills on your desk as the "bank." Write the starting number $155 on the chalkboard.

- One student in each group rolls both number cubes to form a 2-digit number. The number represents the cost of an item. Another student writes the starting amount ($155) and the item's cost in vertical form on a sheet of paper.
- Group members work together to figure out how to pay for the item in exact change. If necessary, they may go to the "bank" at any time and trade one $10 bill for ten $1 bills.
- After paying for the item, the group determines how much money they have left. A group member records the amount on the sheet of paper.
- Group members switch roles and repeat the activity beginning with the starting number.
- Vary the activity by using different starting numbers.

Customizing the Activity for Individual Needs
ESL Help students understand the term *regroup* by modeling regrouping with place-value blocks, place-value charts, and writing examples on the board while explaining the steps and the regrouping symbols.

Learning Styles Students can:

 cross out digits as they regroup and write the renamed digits above them.

 use place-value blocks to help them regroup and subtract by exchanging ten ones blocks for one ten block, and so on.

 count out loud while regrouping and exchanging tens for ones.

Reinforcement Activity
Give students subtraction problems with one regrouping. After students find the difference, ask them to write a related addition problem. For example, students find 263 – 28 = 235. They write 235 + 28 = 263.

Alternative Assessment
Students can demonstrate their understanding of subtraction with regrouping by using $10 and $1 bills in play money.

Example: You have six $10 bills and four $1 bills. How can you pay for a $38 item using exact change? How much will you have left?

Answer: Trade one $10 bill for ten $1-bills. Use three $10 bills and eight $1 bills to pay for the item. That leaves two $10 bills and six $1 bills, or $26.

MATH IN YOUR LIFE
Monthly Expenses
Student Edition page 53

Lesson Objectives
- Apply subtraction to monthly expenses.
- Life Skill: Calculate monthly expenses.
- Life Skill: Find the difference between monthly earnings and monthly expenses.
- Communication Skill: Ask for information over the phone.
- Communication Skill: Explain how to estimate monthly expenses.

Activities
Look It Up
Materials: real estate ads, chart paper, pencils
Procedure: Encourage students to look in newspapers or use the Internet to find the range of rents for different-sized dwellings in your location.

(Continued on p. 30)

(Continued from p. 29)

- Have students make a chart showing the size of the dwelling (1, 2, or 3 bedrooms, and so on.) and the highest and lowest rents advertised.
- Have them calculate the difference between the lowest and highest rents in each category.

Make Phone Calls
Materials: telephone, paper, pencils

Procedure: Let students call 1-800-555-1212 to get the 800 numbers for several long-distance providers.

- Ask students to call different providers to find out how much a 10-minute long-distance call to another state from your city or town would cost under their different plans.
- Have students compare rates and calculate how much they would save using the provider with the lowest rates.

Act It Out
Materials: paper, pencils

Procedure: Give students the following scenario:

Jill makes $895 per month. Her monthly rent is fixed at $375. The total for her phone, utilities, and grocery bills varies from month-to-month.

- Ask students to explain how Jill can estimate her monthly expenses. Have them use the numbers for these expenses from their books to determine the highest and lowest estimates.
- Then have them determine how much money Jill would have left over, if any.
- Ask them to suggest what Jill might do if she wants to buy a new car with monthly payments of $205.

Practice
Have students complete Math in Your Life 31 *Monthly Expenses* from the Classroom Resource Binder.

Internet Connection
Students can use the following Web sites to find out about different long-distance calling plans.

AT&T
   http://www.att.com
Sprint
   http://www.sprint.com/sprint
MCI WorldCom
   http://www.mciworldcom.com

3.6 Subtracting with More Than One Regrouping
Student Edition pages 54–55

Prerequisite Skills
- Subtracting with one regrouping
- Subtracting larger numbers

Lesson Objectives
- Subtract numbers with more than one regrouping.
- Life Skill: Use addition to check subtraction.
- Life Skill: Use addition and subtraction to find passbook balances.

Cooperative Group Activity

Let's Regroup!
Materials: number cubes, paper, pencils, red pencils

Procedure: Organize students into groups of four. Give each group five number cubes. Give each group member a red pencil.

- One student in each group tosses 3–5 number cubes. Group members form a number and write it on their papers.
- Another student tosses 3–5 number cubes again. Group members form a second number and write it on their papers.
- Each student subtracts the smaller number from the larger number. Students use their red pencils to show any regrouping that is needed.
- Group members exchange papers and check each other's work using addition.
- Students repeat the activity several times.

Customizing the Activity for Individual Needs
ESL To help students understand the meaning of *regroup*, have them use base ten blocks to regroup 1 hundred as 10 tens and 1 ten as 10 ones. As they do so, have them explain aloud using the word *regroup*.

Learning Styles Students can:

 write subtraction problems on grid paper to line up digits and write renamed numbers in the squares above the regrouped digits.

 arrange number cards to model subtraction problems, replacing regrouped digits with the appropriate number cards.

30 • Chapter 3

explain each regrouping to a friend or themselves while subtracting and checking answers aloud.

Reteaching Activity

Have students model subtraction with more than one regrouping using different color counters. For example, students can start with 10 blue counters for tens and 20 white counters for ones. They can model 81 using 8 blue counters and 1 white counter. To subtract 25, they can trade one blue counter for 10 white counters, and remove 2 blue and 5 white counters. The remaining number of counters (5 blue and 6 white) models the difference.

Alternative Assessment

Students can identify when to regroup in a subtraction problem.

Example: When would you regroup in this subtraction problem?

$$\begin{array}{r} 23{,}413 \\ -\ 9{,}378 \end{array}$$

Answer: in the ones, tens, hundreds, and thousands places

3.7 Regrouping with Zeros
Student Edition pages 56–57

Prerequisite Skill
- Subtracting with regrouping

Lesson Objectives
- Subtract numbers with one or more zeros.
- Life Skill: Use subtraction to compare numbers in a table.

Cooperative Group Activity

Round-Robin Buying

Materials: index cards, paper, pencils, play money

Procedure: Assign students to groups of four. Give each group 20 index cards, money models for one $100 bill, ten $10 bills, and ten $1 bills.

- Students write a dollar amount that ranges from $11 to $99 on each index card. Cards are placed in a pile. One group member is elected to start, and holds the $100 bill. Remaining bills are placed in the center of the table.

- The starter draws a card, subtracts the amount shown from the $100 bill, and then makes change from the bills in the center. The other students write the problem on paper and find the difference. The starter and the other group members compare to see if they got the same answer.

- The $100 bill is then passed to the member on the starter's left. The other bills are returned to the center of the table, and the activity is repeated. Students continue until all the cards have been used.

Customizing the Activity for Individual Needs

ESL Have students work in pairs to reinforce the term *subtraction with zeros*. One student can show the difference using the play money while the other student verbalizes the difference using the words *subtracting* and *zero*. Students can then switch roles.

Learning Styles Students can:

 write each subtraction problem on the chalkboard.

 use play money to make change and check subtraction by recounting the play money.

 count aloud as play money is used to make change.

Enrichment Activity

Provide each student with a "scoreboard" in the form of an octagon. Have students subtract a 2-digit number from a 2-digit multiple of 10. For example, 40 – 38, 50 – 32, 20 – 13, 60 – 11, 80 – 46, 90 – 12, 70 – 16, 30 – 26. For every correct answer, the student checks off a vertex of the octagon, signifying a correct answer. Continue until all students have completed a circuit of their personal scoreboard.

Alternative Assessment

Students can explain how to regroup a number that ends in one or more zeros.

Example: How can you regroup 100 to subtract?

Answer: Regroup the 1 hundred to make 10 tens, then regroup 1 ten to make 10 ones. You will then have 9 tens and 10 ones to subtract.

3.8 Problem Solving: Add or Subtract
Student Edition pages 58–59

Prerequisite Skills
- Knowing clue words for addition
- Knowing clue words for subtraction

Lesson Objectives
- Decide whether to add or subtract to solve a word problem.
- Life Skill: Draw a diagram to solve a problem.

Cooperative Group Activity

Addition and Subtraction Trivia
Materials: almanacs, self-stick notes, pencils, paper

Procedure: Use self-stick notes to mark several sections in each almanac; for example, number of animals at different zoos or the populations of different countries. The numbers should be whole numbers. Organize students into teams. Give each team three index cards and a marked almanac. Assign one marked section to each team.

- Each team writes three questions on paper that can be answered by adding or subtracting the information in the marked section of their almanac.
- Teams trade questions. Team members work together to find the information they need in the almanacs. Then they do the calculations to answer the questions and check their work.

Customizing the Activity for Individual Needs
ESL Review with students the clue words for addition and subtraction. They are listed on pages 32 and 48 of the Student Edition. Have students write + or – as they read the various clue words. Remind students that for whole numbers, addition results in larger numbers; subtraction results in smaller numbers.

Learning Styles Students can:

 circle addition clue words and underline subtraction clue words in exercises as they solve problems.

 write a + sign on one index card and a – sign on another index card, then hold up the correct sign to indicate which operation to use.

 check their work aloud and explain whether answers make sense.

Reinforcement Activity

Provide students with a large quantity of play money. Each student starts with a $100 bill and the rest is placed in the center of a table. One student designates "subtraction" or "addition" while giving a word problem that illustrates that operation. Students decide whether the word problem is correctly solvable by the designated operation. They use the money to subtract or add from the money in the center, regrouping when needed.

Alternative Assessment

Students can identify words or phrases that indicate subtraction and addition.

Example: Which words in the list tell you to subtract? Which words tell you to add?

both	how much more
altogether	how much less
left	remain
in all	how many more
difference	total

Answer: Subtract: left, remain, difference, how much more, how much less, how many more. Add: both, altogether, in all, total.

Closing the Chapter
Student Edition pages 60–61

Chapter Vocabulary
Review with the students the Words to Know on page 43 of the Student Edition. Then have students quiz each other in pairs.

Have students copy and complete the Vocabulary Review questions on page 60 of the Student Edition.

For more vocabulary practice, have them complete Vocabulary 23 *Subtracting Whole Numbers* from the Classroom Resource Binder.

Test Tips
In pairs, students can practice writing subtraction problems in vertical form using the test tip idea about lining up digits by place value.

Learning Objectives
Have the students review *Goals and Self-Check* CM6. They can check off the goal they have reached. Note that each section of the quiz corresponds to a Learning Objective.

Group Activity
Summary: Students look through newspapers to compare prices and features of the same model car from different places.

Materials: newspapers, pencils, paper

Procedure: Students can make a chart to help them compare the prices and features of the same model car. Students then choose the best car to buy.

Assessment: Use the Group Activity Rubric on page xii of this guide. Fill in the rubric with the additional information below. For this project, students should have:
- compared the features of a least two cars.
- compared the prices of at least two cars.

RELATED MATERIALS See the unit overview page for other Globe Fearon books that can be used to enrich and extend the material in this unit.

Assessing the Chapter

Traditional Assessment
Chapter Quiz
The Chapter Quiz on pages 60–61 of the Student Edition can be used as either an open- or closed-book test, or as homework. The quiz can be used to identify concepts in the chapter that students need to review and practice.

Chapter Tests
Use Chapter Test A Exercise 33 and Chapter Test B Exercise 34 *Subtracting Whole Numbers* from the Classroom Resource Binder to further assess mastery of chapter concepts.

Additional Resources
Use the Resource Planner on page 24 of this guide to assign additional exercises from the Classroom Resource Binder and Workbook.

Alternative Assessment
Interview
Write the numbers 4,712 and 2,965 on a piece of paper. Ask:
- Which is the larger number?
- How could you find the difference between these numbers?
- What operation would you use to find the difference?
- How would you write these numbers to help you subtract?
- How would you subtract the smaller number from the larger one? Where did you use regrouping?

Real-Life Connection
Have students find a newspaper article that contains at least one large number. Then instruct students to:
- make up a subtraction problem that uses the number(s) found.
- solve the problem.

Planning the Chapter

Chapter 4 • Multiplying Whole Numbers

Chapter at a Glance

SE page	Lesson	
62		Chapter Opener and Project
64	4.1	What Is Multiplication?
65	4.2	Basic Multiplication
66	4.3	Multiplying Larger Numbers
69		Using Your Calculator: Checking Multiplication
70	4.4	Multiplying with One Regrouping
72	4.5	Multiplying with More Than One Regrouping
74	4.6	Problem Solving: Clue Words for Multiplication
76	4.7	Multiplying Whole Numbers by 10, 100, 1,000
77		On-The-Job Math: Inventory Clerk
78	4.8	Multiplying by Numbers That Contain Zero
80	4.9	Problem Solving: Two-Part Problems
82		Chapter Review and Group Activity

Learning Objectives
- Multiply whole numbers.
- Multiply larger numbers.
- Multiply with regrouping.
- Multiply numbers by 10, 100, 1,000.
- Multiply by numbers that contain zero.
- Solve word problems using multiplication.
- Solve two-part word problems.
- Apply multiplication to counting items for inventory.

Life Skills
- Multiply to find costs.
- Use a calculator to multiply and check multiplication.
- Plan and budget for a vacation.
- Find distances traveled.
- Work backward to solve a problem.
- Compare car mileages.

Communication Skills
- Read nutrition labels.
- Write a letter.
- Explain how to take inventory.

Workplace Skills
- Take inventory.
- Estimate total receipts.

PREREQUISITE SKILLS
To assess mastery of prerequisite skills, use a selection of exercises from any of the program resources referenced below. The same resources can be used to provide remediation if necessary.

| Skills | Program Resources for Review | | |
	Student Edition	Workbook	Classroom Resource Binder
Adding one-digit numbers	Lessons 2.2, 2.3	Exercises 11, 12	
Adding larger numbers	Lessons 2.4, 2.6, 2.7	Exercises 13, 15, 16	Review 14, 15 Practice 17

Diagnostic and Placement Guide

Chapter 4 Multiplying Whole Numbers

The Percent Accuracy scores are based on the number of problems in a lesson that have been answered correctly.

Resource Planner

After diagnosing your students' needs, use the Correlating Program Resources for reinforcement, reteaching, or enrichment. Additional activities for customizing lessons can be found in this guide.

KEY
Reteaching = ↶   Reinforcement = ↓   Enrichment = ↷

Lessons		Percent Accuracy				Program Resources			
		50%	65%	80%	100%	↓ Student Edition • Extra Practice	↓ Workbook • Exercises	Teacher's Planning Guide	Classroom Resource Binder
4.1	What Is Multiplication?	5	7	8	10		26	↶ p. 36	
4.2	Basic Multiplication	8	10	13	16	p. 418	27	↶ p. 37	
4.3	Multiplying Larger Numbers	17	22	27	34	p. 418	28	↶ p. 38	
4.4	Multiplying with One Regrouping	17	22	27	34	p. 418	29	↓ p. 39	↓ Review 36 Multiplying with One Regrouping ↓ Practice 39 Multiplying with One Regrouping
4.5	Multiplying with More Than One Regrouping	9	12	14	18	p. 418	30	↓ p. 40	↓ Review 37 Multiplying with More Than One Regrouping ↓ Practice 40 Multiplying with More Than One Regrouping
4.6	Problem Solving: Clue Words for Multiplication	1	2	2	3		31	↓ p. 41	↓ Practice 41 Problem Solving: Clue Words for Multiplication
4.7	Multiplying Whole Numbers by 10, 100, 1,000	6	8	10	12		32	↶ p. 42	↓ Mixed Practice 44 Adding, Subtracting, and Multiplying Whole Numbers
	On-The-Job Math: Inventory Clerk	Can be used for portfolio assessment.						↷ p. 42	↷ On-The-Job Math 45 Inventory Clerk
4.8	Multiplying by Numbers That Contain Zero	11	14	17	21		33	↓ p. 43	↓ Review 38 Multiplying by Numbers That Contain Zero ↓ Practice 42 Multiplying by Numbers That Contain Zero ↓ Mixed Practice 44 Adding, Subtracting, and Multiplying Whole Numbers
4.9	Problem Solving: Two-Part Problems	1	2	2	3		34	↶ p. 44	↓ Practice 43 Problem Solving: Two-Part Problems
Chapter 4 Review									
	Vocabulary Review	4	5.5 (writing is worth 4 points)	6.5	8			↓ p. 45	↓ Vocabulary 35
	Chapter Quiz	15	20	24	30			p. 45	↓ Chapter Test A, B 47–48

Customizing the Chapter

Opening the Chapter
Student Edition pages 62–63

Photo Activity
Discuss rate × time = distance. Point out that the speed of sound is measured in *feet per second*, while other speeds, such as that of a car, are measured in *miles per hour*. Using a rate of 5 miles per hour, have students figure out how much distance is covered in 2 hours, 3 hours, . . . Notice that the distance can be found by skip counting or multiplying the number of hours by 5 (the rate). Present the following data on the maximum animal speeds: lion, 50 mph; grizzly bear, 30 mph; human, 27 mph; black mamba snake, 20 mph. Ask: *What is the farthest a lion could travel in 2 hours? What is the farthest a snake could travel in 5 hours? How far can a human go in 5 hours compared with a grizzly bear?*

Words to Know
Review with the students the Words to Know on page 63 of the Student Edition. Help students remember these terms by having them make a concept map. Suggest that they start with the heading "multiplication" and draw arrows between related terms. For example:

factors → partial product → product.

The following words and definitions are covered in this chapter.

multiplication a quick way to add; repeated addition

multiply to add a number to itself one or more times; 2 + 2 + 2 + 2 = 8 or 4 × 2 = 8

factors the numbers that are multiplied to obtain a product

product the final answer to a multiplication problem

partial product number obtained by multiplying a number by only one digit of a two or more digit number

Nutrition Label Project
Summary: Students collect nutrition labels, then multiply by the total number of servings to find the total number of Calories in an entire package.

Materials: nutrition labels from various foods, paper, pencils, tape or glue

Procedure: Students make scrapbooks of their nutrition labels with their calculations. They can add more labels to their scrapbooks as they come across other foods. Scrapbooks can be made in class with materials provided, then calculations done as homework. To extend this project, students can create their own labels that show nutritional information for foods that are not usually packaged or for made-up sizes and combinations.

Assessment: Use the Individual Activity Rubric on page *xi* of this guide. Fill in the rubric with the additional information below. For this project, students should have:

- collected at least three nutrition labels.
- multiplied correctly the number of servings in a package by the number of calories per serving.

Learning Objectives
Review with the students the Learning Objectives on page 63 of the Student Edition before starting the chapter. Students can use the list of objectives as a learning guide. Suggest that they write the objectives in a journal or use *Goals and Self-Check* CM6.

After each lesson, have students write an example of the skill they learned under the appropriate objective. Suggest that students use these notes as a learning guide to help them study for the chapter test.

4.1 What Is Multiplication?
Student Edition page 64

Prerequisite Skill
- Adding one-digit numbers

Lesson Objectives
- Write multiplication problems in horizontal and vertical form using numbers.
- Change repeated addition to multiplication.

Words to Know
multiplication, multiply, factors, product

Cooperative Group Activity

Making Rectangle Products
Materials: large-squared grid paper or Visual 12 *Hundredths Model*, markers or crayons, pencils

36 • Chapter 4

Procedure: Organize students into groups of four. Demonstrate how to find products using rectangles. Draw a grid on the chalkboard. Color in a 2-by-3 rectangle. Point out that the rectangle contains 2 groups of 3 squares, with 6 squares in all. Write the multiplication problem 2 × 3 = 6. Also show students a 3-by-2 rectangle, to represent 3 groups of 2, or 3 × 2 = 6. Give each student a copy of Visual 12 or a sheet of grid paper.

- Each student in the group colors in some squares in the grid to form any size and shape rectangle. This represents the product.
- Group members exchange grids. Students determine the multiplication problem that corresponds to the number of rows, columns, and squares forming the rectangle.
- Students repeat the process with blank grid paper and a new multiplication fact. Encourage them to make at least four different facts.
- Challenge students to write a multiplication problem on blank grid paper and have others then color the appropriate rectangle.

Customizing the Activity for Individual Needs

ESL Help students differentiate between everyday and mathematical meanings of the words *multiply*, *factors*, and *product*. Give examples of everyday meanings, then model the mathematical definitions in context. For example: One *factor* in her decision to go running was the weather. The numbers you multiply are called the *factors*.

Learning Styles Students can:

 draw pictures to represent multiplication problems.

 manipulate counters or ones blocks to represent multiplication problems.

 skip count with repeated additions to find products.

Reteaching Activity

Provide each student with two copies of Visual 1 *Number Lines: Whole Numbers* or have students draw number lines. Show students how to make "jumps" on a number line to find products. For example, to find 3 × 6, students start at 0 and make 3 "jumps" of 6 units each.

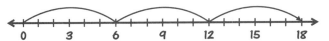

Alternative Assessment

Students can show multiplication problems by drawing a picture.

Example: Draw a picture to show 3 × 5 = 15.

Answer: Student draws 3 groups of 5 objects connected by plus signs and one group of 15 objects.

4.2 Basic Multiplication
Student Edition page 65

Prerequisite Skills
- Understanding multiplication
- Reading tables

Lesson Objective
- Use a multiplication table to find basic facts.

Cooperative Group Activity

Find the Fact

Materials: 10-by-10 grid paper or Visual 12 *Hundredths Model*, small counters, paper, pencils

Procedure: Organize students into pairs. Give one copy of grid paper or Visual 12 and a counter to each pair.

- Partners fill in Visual 12 to create their own multiplication tables as shown below. Factors should be written outside the grid, along the top and side as shown below. Products should be written inside the grid.

×	1	2	3	4	5	6	7	8	9	10
1	1	2	3	4	5	6	7	8	9	10
2	2	4	6	8	10	12	14	16	18	20
3	3	6	9	12	15	18	21	24	27	30
4	4	8	12	16	20	24				
5										
6										
7										
8										
9										
10										

- One student in each pair tosses the counter onto the multiplication table. The other student writes a multiplication fact for the product the counter lands on. Then that number is crossed off.
- Partners take turns tossing and writing facts until all products from the table are used.

Customizing the Activity for Individual Needs

ESL To be sure students understand parts of a table and how to use it, define *row* and *column*. Have partners work together reading a table. They can take turns pointing to numbers in the table and naming the corresponding row and column.

Learning Styles Students can:

 describe patterns in the table to help find products. For example, every number in the 5 column has a factor of 5.

 manipulate counters or markers on a table or grid to show the multiplication facts.

 if possible, listen to a prerecorded audiotape of multiplication facts.

Enrichment Activity

Have students present at least two different ways to organize the basic multiplication facts. For example, they can organize them in pairs (commutative property), such as $2 \times 3 = 6$ and $3 \times 2 = 6$; or by starting with the same first factor (multiples), such as $4 \times 1 = 4$, $4 \times 2 = 8$, $4 \times 3 = 12$, and so on. They can also group multiplication facts that have the same product together, such as $1 \times 12 = 12$, $2 \times 6 = 12$, $3 \times 4 = 12$, $4 \times 3 = 12$, and so on. Encourage creative presentation methods such as posters or booklets.

Alternative Assessment

Students can show mastery of basic facts by responding to flash cards.

Example: Show 9×7 on a flash card.

Answer: 63

4.3 Multiplying Larger Numbers
Student Edition pages 66–68

Prerequisite Skills
- Understanding place value
- Adding larger numbers
- Knowing basic multiplication facts

Lesson Objectives
- Multiply larger numbers without regrouping.
- Life Skill: Find the cost of multiple tickets items.

Words to Know
partial product

Cooperative Group Activity

The Play's the Thing

Materials: cardboard or poster board, markers, index cards, money models for $10s and $1s or Visual 15 *Money Models*, pencils, paper

Procedure: Organize students into groups of four. Give each group a sheet of cardboard or poster board, a marker, 10 index cards, twenty $10 bills, and twenty $1 bills. Have groups create signs using the markers and poster board showing ticket prices for a school play. For example, Adults $4; Students and Seniors $2. Prepare index cards as "tickets."

- One student gets the "tickets" and acts as the box office attendant to sell them. Another decides how many and what kind to purchase for his or her family.
- The third student does the multiplication and addition. The fourth student checks the multiplication and addition.
- The second student then uses the play money to purchase the tickets from the attendant.
- Students start over, rotating roles until all purchase tickets for their families.

Customizing the Activity for Individual Needs

ESL Point out that a *partial* product is *part* of a product. Help students identify the partial products of 532×111.

Answer:
$$\begin{array}{r} 532 \\ \times\,111 \\ \hline 532 \\ 5\,320 \\ +\,53\,200 \\ \hline 59{,}052 \end{array}\Bigg\}\text{Partial products}$$

Learning Styles Students can:

 draw vertical lines and use end zeros as placeholders to help align digits.

 model problems with number cards, aligning the cards in their proper places and using zero cards as placeholders.

 repeat the factors and partial products as a partner says them first.

Reteaching Activity

Provide lined paper turned sideways. Have students write problems in vertical form and find the products.

For example, 72×3; 812×4; 923×12; 521×432.

Answers: 216; 3,248; 11,076; 225,072

Alternative Assessment

Students can find an error in multiplication and correct it.

Example: Where is the error in this problem?

```
   521
 × 12
 1 142
+ 5 210
 6,352
```

Answer: First partial product should be 1,042, resulting in a product of 6,252.

USING YOUR CALCULATOR
Checking Multiplication
Student Edition page 69

Lesson Objectives
- Check a multiplication problem.
- Life Skill: Use a calculator to multiply.

Game
Reach 5,000
Materials: calculators, score pads
Procedure: Have students work in pairs to play this game. As students play, encourage them to use mental math to try to outwit their opponent.
- Player 1 enters any 3-digit number on the calculator. Player 2 multiplies this by a 1-digit number other than 0 or 1. (*Note:* Some calculators will require students to press the equals key to display the product.)
- Without clearing, players continue entering digits other than 0 or 1 until the product goes over 5,000. The first player who makes the product go over 5,000 loses that round. The winner scores one point. The first to get 10 points wins the game.

```
Player 1   [2][5][6]     [   256]
Player 2   [×][4][=]     [  1024]

Player 1   [×][2][=]     [  2048]
Player 2   [×][3][=]     [  6144]  OVER!
```

Variations: Change the target to 10,000 or 1,000. Players may pass a turn if they think the product will go over. The first player to go over the target wins the round.

4.4 Multiplying with One Regrouping *Student Edition*
pages 70–71

Prerequisite Skills
- Renaming using place value
- Multiplying larger numbers

Lesson Objectives
- Multiply numbers with one regrouping.
- Life Skill: Find the total cost of a vacation.

Cooperative Group Activity

Spin, Roll, and Multiply
Materials: number cubes with 1–6, place value charts or Visual 2 *Place-Value Charts: Whole Numbers*, blank three-sectioned spinners or Visual 17 *Spinners*, pencils, stopwatches or timers.

Procedure: Organize students into groups of four. Give each group a number cube, copies of the place-value chart, a spinner, and a timer. Prepare the spinner with the following numbers on it: 96; 645; and 4,728. Assign roles: spinner, roller, recorder, timer. Groups compete to be the fastest to spin, roll, and multiply.
- Timer says "Go" and starts the timer.
- Spinner spins and calls out the first factor. Roller rolls and calls out the second factor.
- Recorder writes the multiplication problem (formed by the two factors) in the place-value chart. Together, the group finds the product.
- Timer stops the stopwatch and notes the time.
- Recorder shares their problem and time with the class. Fastest group with a correct product gets a prize or point. (Four points wins a game.)
- Rotate roles and repeat until everyone has played each role or a team wins. (*Note:* If the same problem is formed, the roller rolls again.)

Customizing the Activity for Individual Needs

ESL Pair students to review *renaming* and basic multiplication facts for 5, 6, 7, 8, and 9. One partner gives a multiplication fact and the other renames the product using *tens* and *ones*. For example: 7 x 7 = 49; 49 is 4 tens and 9 ones.

Learning Styles Students can:

 write the carried digits in a different color.

 use place-value blocks to help them multiply and regroup.

 listen as partners say aloud place value of each digit, rename, and regroup as needed.

Enrichment Activity

Provide students with $1, $10, and $100 bills to equal $550. Use Visual 15 *Money Models*. Explain that a one-week hiking vacation at Redding Lodge costs $75 per person. If a canoe trip is included, the cost is $95 per person. Have them model the cost for two people to hike for one week, then five weeks. Have them model the cost of a canoe trip for one person for one week and five weeks.

Alternative Assessment

Students can model multiplication with regrouping by using place-value blocks.

Example: Find 27 × 3 using place-value blocks.

Answer: 81; 3 groups of 2 tens blocks and 7 ones are combined for a total of 6 tens blocks and 21 ones blocks. Trade 20 ones for 2 tens for a total of 8 tens blocks and 1 ones block.

4.5 Multiplying with More Than One Regrouping
Student Edition pages 72–73

Prerequisite Skills
- Multiplying with one regrouping
- Multiplying larger numbers

Lesson Objectives
- Multiply numbers with more than one regrouping.
- Life Skill: Find distance, given the speed and time traveled.

Cooperative Group Activity

Number Cube Problems

Materials: paper, pencils, number cubes

Procedure: Organize students into pairs. Give each pair four number cubes, numbered from 1–6.
- One partner rolls three number cubes to form a 3-digit number. Both students record the number.
- The other student rolls two number cubes to form a 2-digit number. Both students record the number.
- Pairs work together to find the product of the two numbers.
- Students can vary the activity to include factors with 2, 3, and 4 digits.

Customizing the Activity for Individual Needs

ESL Help students understand the term *separate regroupings* by breaking down each problem into parts. Have students first find each partial product separately. Highlight this with a marker. Then find their sum. Students can then do the whole problem. They see where each piece fits.

Example: Multiply 536 × 96.

$$\begin{array}{r} {\scriptstyle 2\,3} \\ 536 \\ \times\ 6 \\ \hline 3{,}216 \end{array} \quad \begin{array}{r} {\scriptstyle 3\,5} \\ 536 \\ \times\ 90 \\ \hline 48{,}240 \end{array} \quad \begin{array}{r} {\scriptstyle 1} \\ 3\ 216 \\ +\ 48\ 240 \\ \hline 51{,}456 \end{array} \quad \begin{array}{r} {\scriptstyle 3\,5} \\ {\scriptstyle 2\,3} \\ 536 \\ \times\ 96 \\ \hline {\scriptstyle 1}3\ 216 \\ +\ 48\ 240 \\ \hline 51{,}456 \end{array}$$

Learning Styles Students can:

 use two different colors for each row of carried numbers.

 arrange number cards to model multiplication problems, replacing regrouped digits with the appropriate number cards.

 take turns explaining each regrouping to a small group while multiplying.

Reinforcement Activity

Copy the table below. Then have students complete the table by determining the distance traveled for each biker. Ask students to identify the multiplication problems in which they had to regroup.

Biking Club Contest			
Biker	Average Speed	Hours Biked	Total Distance
Ann	35 mph	14	
Alex	29 mph	25	
Kyle	25 mph	18	
Jane	36 mph	17	

Answers: Ann, 490 miles; Alex, 725; Kyle, 450; Jane, 612; all require regrouping.

Alternative Assessment

Students can fill in the missing regrouping digits in a multiplication problem.

Example: Fill in the regrouping digits in this multiplication problem.

$$\begin{array}{r}\square\square\\ \square\square\\ 375\\ \times\ 23\\ \hline 1\,125\\ +7\,500\\ \hline 8{,}625\end{array}$$

Answers: top left box, 1, top right, 1; bottom left, 2, bottom right, 1

4.6 Problem Solving: Clue Words for Multiplication
Student Edition pages 74–75

Prerequisite Skills
- Knowing basic multiplication facts
- Multiplying larger numbers

Lesson Objectives
- Recognize clue words for multiplication.
- Solve word problems using multiplication.
- Work backward to solve a problem.
- Life Skill: Find multiple unit pricing.

Cooperative Group Activity

Numbers, Clues, Numbers

Materials: four-part blank spinner or Visual 17 *Spinners*, paper, pencils, index cards, number cubes, number cards 1 to 100 only or Visual 4 *Whole Numbers 1 to 100* cut into cards

Procedure: Organize students into pairs. Make spinners with the following clue word phrases: in groups of, for $... each, in... hours, at $... each.

Give each pair three index cards, two number cubes, a spinner, and number cards 1 to 100.

Write the following parts on the chalkboard.

Parts for Word Problems
first number
item(s)
clue word(s)
second number

- One partner rolls a number cube to find the first number and states what the number represents.
Example: 5 comic books
- The other partner spins to get the clue words. He or she selects a number card to form the second part of the word problem.
Example: at $16 each
- Pairs use the "parts" to write a word problem on the front of the index card. The solution should be written on the back of the card.
Example: I bought 5 comic books at $16 each.
Solution: 5 × $16 = $80
- Repeat to complete two more index cards. When all pairs have finished writing their problems, a volunteer collects the cards. The cards are placed problem-side up in a pile in the front of the room.
- Each pair picks a card from the pile without looking. They solve the problem together by identifying the clue words, then showing their multiplication work. They check their answers by looking at the back of the index cards.

Customizing the Activity for Individual Needs

ESL Pairs of students help each other read word problems and identify the clue words. They take turns, one stating the embedded multiplication problem and the other solving it.

Learning Styles Students can:

 draw pictures to illustrate and solve the multiplication word problems.

 use real-life objects to model the multiplication word problems.

 listen as partners read aloud problems and identify multiplication clue words. They then say aloud the multiplication problem to be solved.

Reinforcement Activity

Have students pick a word problem from a textbook, workbook, or worksheet. (See p. 35 for specific references.) Have them make a poster illustrating the real-world situation and how to solve it using the four problem-solving steps. Use Visual 22 *Problem-Solving Steps*.

Alternative Assessment

Students can identify clue words in a problem that indicate multiplication.

Example: What clue words in this problem tell you to multiply?

Joe bought 2 video games at $26 each. How much did he spend?

Answer: at $26 each

4.7 Multiplying Whole Numbers by 10, 100, 1000
Student Edition page 76

Prerequisite Skill
- Understanding place value

Lesson Objective
- Multiply numbers by 10, 100, and 1,000.

Cooperative Group Activity

10, 100, and More

Materials: paper, pencils

Procedure: Assign students to groups of four.
- One student picks a number less than 99,999.
- Each of the other group members multiplies the number by 10, by 100, and by 1,000 and records the results on paper.
- Students take turns choosing the starting number. At the chalkboard, a volunteer from each group explains one of the multiplication problems the group has completed.

Customizing the Activity for Individual Needs

ESL Help students understand that a *shortcut* is a way of saving time. To illustrate, have students find 831 × 100, doing the multiplication the long way using partial products. Then demonstrate the short way by writing zeros.

Learning Styles Students can:

highlight zeros in the power of ten factor and the same number of zeros in the product.

cut out the zeros from the power of ten factor from the multiplication problem. They then can rewrite the number and move the zeros into place to the right of the number.

says the number of zeros in the factor and the product aloud.

Reinforcement Activity

Have students multiply, using multiples of 10, 100, and 1,000 as one of the factors. Provide grid paper. Students write a problem such as 3,000 × 60 in one row of boxes near the right-hand side of the grid. They circle all the zeros and write the same number of zeros in the row below, starting at the far right. They multiply the remaining digits to complete the answer. Pairs can make up problems for partners to multiply.

Alternative Assessment

Students can show their understanding of multiplying by 10, 100, or 1,000 by matching multiplication problems with their products.

Example: Match each problem in the first row with its product in the second row.

1. 39 × 10   2. 39 × 100   3. 39 × 1,000
(a) 3,900   (b) 39,000   (c) 390

Answers: 1–c; 2–a; 3–b

ON-THE-JOB MATH
Inventory Clerk
Student Edition page 77

Lesson Objectives
- Apply multiplication to counting items for inventory.
- Communication Skill: Write a letter requesting an employment application.
- Workplace Skill: Create an inventory table.

Activities

Tools of the Trade

Materials: inventory table, pencils

Procedure: Have small groups work together to inventory common classroom items. As a class, discuss items to be inventoried, such as chalk, pencils, and paper supplies. They can use an inventory table like the one shown below. Remind them to multiply to complete the last column. Meet as a class to discuss and share findings.

Name of Item	Number of Containers	Amount per Container	Total Amount
chalk	3 boxes	12	36

Write a Business Letter
Materials: telephone books, paper, pencils
Procedure: Have students use local telephone books to find addresses for several supermarkets, department stores, discount stores, or other businesses that sell goods. Have them write letters to ask:
- How often is inventory taken?
- Is inventory done by a store employee or an outside firm?
- What methods are used in taking inventory?
- Are there employment applications or other information available about applying for an inventory job?

When responses to students' letters come in, use the information in a whole-class activity to compile the results into a chart.

Act It Out
Materials: none
Procedure: Have students role-play being inventory clerks who are responsible for taking an accurate inventory as of noon every Monday. Explain:
- While you are taking inventory, the store manager needs to remove 2 boxes of light bulbs from the storage room. The light bulbs have not yet been counted in the inventory. What do you say to the manager?

Have students offer suggestions for ways to satisfy the manager's needs and also keep their inventory accurate.

Practice
Have students complete On-The-Job Math 45 *Inventory Clerk* from the Classroom Resource Binder.

Internet Connection
The following Web sites are starting points for researching employment opportunities in inventory, writing résumés, and other career-related information.

Career Mosaic
    http://www.careermosaic.com
The Internet's Online Career Center
    http://occ.com

4.8 Multiplying by Numbers That Contain Zero
Student Edition pages 78–79

Prerequisite Skill
- Multiplying with regrouping

Lesson Objectives
- Multiply numbers that have zero as a digit.
- Life Skill: Compare distances that cars can travel.

Cooperative Group Activity

Zero Is in on It
Materials: number cards labeled 0–9, paper, pencils

Procedure: Have students work in pairs. Give each pair a set of number cards and ask them to place the zero on the table face up. Have them turn the remaining cards face down.
- Students try to see what effect zero has in the ones place or hundreds place of a 3-digit factor. One partner draws three cards. The student uses those cards, along with the 0 card, to make a 3-digit number times a 1-digit number multiplication problem.
- Partners work together to solve the problem. Then they exchange roles and repeat the activity with the zero in a different position.
- Students then describe what would happen if zero were used as the 1-digit factor.

Customizing the Activity for Individual Needs
ESL To reinforce the term *zero property* of multiplication, use Visual 5 *Base Ten Blocks* to demonstrate multiplication. Say, "3 groups of 10 is 30" as you show three tens. Say, "1 group of 10 is ten" as you show one ten. Say, "0 groups of 10 is zero" as you show and gesture nothing.

Learning Styles Students can:

write problems on lined paper turned sideways so that digits can be kept in line.

rearrange number cards to place zero in various positions.

say each step to themselves as they write partial products that are zero.

Reinforcement Activity

Copy the table below. Then have students use the table to answer the following questions: How far can each model go on a full tank of gas? Which car(s) can travel 1,100 miles on a full tank of gas?

New Models Mileage		
Car	Miles per Gallon	Tank Size
X	50 mpg	20 gal
Y	46 mpg	25 gal
Z	48 mpg	24 gal

Answers: X, 1,000; Y, 1,150; Z, 1,152; Y and Z

Alternative Assessment

Students can identify the mistake in a multiplication problem.

Example: What mistake was made in this multiplication problem? What is the correct answer?

```
     527
  × 205
  ─────
   2 635
 +10 540
  ──────
  13,175
```

Answers: The second partial product (527 × 0) was left out. The third partial product should be 105,400. The correct answer is 108,035.

4.9 Problem Solving: Two-Part Problems *Student Edition pages 80–81*

Prerequisite Skills
- Adding, subtracting, and multiplying whole numbers
- Identifying clue words for addition, subtraction, and multiplication

Lesson Objectives
- Solve two-part word problems.
- Make a table to solve a problem.
- Life Skill: Find total costs.

Cooperative Group Activity

Placing a Sales Order
Materials: sales catalogs, number cubes, pencils, paper
Optional: Visual 16 *Blank Table*

Procedure: Have students work in pairs. Distribute a different sales catalog and a number cube to each. Give students a budget of $500.
- Pairs create a four-column chart with the following headings: Item, Quantity, Price of Item, and Total.
- Pairs choose items from the catalog to order. They enter the name of each item and the item price in their charts, ignoring any decimal parts.
- Partners take turns tossing the number cube to determine the quantity of each item ordered.
- Pairs find the total price for each item ordered and the cost of their entire order. They cannot exceed $500.

Variation: Change the budget amount. Allow calculators.

Customizing the Activity for Individual Needs

ESL Review the clue words for addition, subtraction, and multiplication. Students use a highlighter to mark the clue words in the problems as they say them and write the symbol for the appropriate operation above each clue.

Learning Styles Students can:

 use different colors to highlight clue words for different operations and other important phrases in word problems, then make a key for the colors and operations they represent.

 write +, −, and × signs on separate index cards, then hold up a sign to indicate which operation to use for a part of a problem.

 explain why an answer a partner gives does or does not make sense.

Enrichment Activity

Have students fill in cashiers' *Cash Register* tables. Provide play money or Visual 15 *Money Models* and Visual 16 *Blank Table*. Sample "cash drawer": 32 $10 bills, 27 $5 bills, and 14 $1 bills. Have pairs take turns making "money draws" and "cashing out."

Cash Register Table		
Bills	Count	Amount
$10	32	$320
$5	27	$54
$1	14	$14

Alternative Assessment

Students can draw a diagram of a plan to solve a two-part problem.

Example: Draw a diagram for the problem below.

   Bill bought 5 lamps. Each lamp cost $40. He also bought light bulbs for $10. How much did Bill spend in all?

Sample answer: 5 lamps each with a $40 tag; sales slip says 5 × $40 = $200; light bulbs with a $10 tag. A ring around items shows "adding."

Closing the Chapter
Student Edition pages 82–83

Chapter Vocabulary
Review with the students the Words to Know on page 63 of the Student Edition. Then have students quiz each other in pairs.

Have students copy and complete the Vocabulary Review questions on page 82 of the Student Edition. For more practice, use Vocabulary 35 *Adding Whole Numbers* from the Classroom Resource Binder.

Test Tips
Have students read each test tip, then pick out problems in which they can apply that tip. Encourage students to explain how the tip will help them with a particular type of problem.

Learning Objectives
Have the students review *Goals and Self-Check* CM6. They can check off the goal they have reached. Note, each quiz section corresponds to a Learning Objective.

Group Activity
Summary: Students calculate stadium receipts for a concert tour of four cities.

Materials: almanacs, newspaper ads for concerts, pencils, paper

Procedure: Encourage students to name their concert tour. Have them take turns choosing locations and stadiums. They may wish to charge different amounts at different stadiums.

Assessment: Use the Group Activity Rubric on page xii of this guide. Fill in the rubric with the additional information below. For this project, students should have:
- found seating capacities for four stadiums.
- decided on ticket prices and calculated estimates based on all performances.

RELATED MATERIALS See the unit overview page for other Globe Fearon books that can be used to enrich and extend the material in this unit.

Assessing the Chapter

Traditional Assessment
Chapter Quiz
The Chapter Quiz on pages 82–83 of the Student Edition can be used as either an open- or closed-book test, or as homework. The quiz can be used to identify concepts in the chapter that students need to review and practice.

Chapter Tests
Use Chapter Test A, Exercise 47 and Chapter Test B, Exercise 48, *Multiplying Whole Numbers* from the Classroom Resource Binder to further assess mastery of chapter concepts.

Additional Resources
Use the Resource Planner on page 35 of this guide to assign additional exercises from the program.

Alternative Assessment
Interview
Write the problem 19 × 407 on a piece of paper. Ask:
- What multiplication facts do you need to find the product?
- How many partial products will there be? How do you know?
- What is the product? *Answer:* 19 × 407 = 7,733
- How could you use a calculator to check your multiplication?
- Four hundred seven people bought tickets for the zoo. Each ticket cost $19. How is this word problem related to the product you found?

Portfolio
Provide students with order forms similar to those found in mail-order catalogs. Have students:
- choose at least three items to order
- order more than one of each item
- find the total cost of each item ordered and the total cost of the order
- write a short explanation of how multiplication is needed to complete an order form

Planning the Chapter

Chapter 5 • Dividing Whole Numbers

Chapter at a Glance

SE page	Lesson	
84		*Chapter Opener and Project*
86	5.1	What Is Division?
87	5.2	Basic Division
88	5.3	Dividing with Remainders
90	5.4	Dividing Larger Numbers
92	5.5	Checking Division
93		*Using Your Calculator: Checking Division*
94	5.6	Problem Solving: Clue Words for Division
96	5.7	Dividing by Numbers with More Than One Digit
98	5.8	Zeros in the Quotient
100	5.9	Problem Solving: Choose the Operation
102		*Math in Your Life: Determining Miles per Gallon (mpg)*
103	5.10	Estimating and Thinking
104		*Chapter Review and Group Activity*

Learning Objectives
- Divide whole numbers.
- Divide larger numbers.
- Divide and get remainders.
- Check division problems.
- Use estimating to choose the best answer.
- Solve word problems using division.
- Solve word problems using any operation.
- Apply division to find miles per gallon.

Life Skills
- Apply division to seating arrangements.
- Calculate equal amounts and payments.
- Use a calculator, check division and other operations.
- Use a formula to find rate and time.
- Compare gas mileages.
- Use a dictionary.
- Understand unit pricing.

Communication Skills
- Write ways division is used in daily life.
- Explain how to find miles per gallon.
- Write a letter requesting brochures.
- Negotiate for the best price.

PREREQUISITE SKILLS
To assess mastery of prerequisite skills, use a selection of exercises from any of the program resources referenced below. The same resources can be used to provide remediation if necessary.

	Program Resources for Review		
Skills	Student Edition	Workbook	Classroom Resource Binder
Adding whole numbers	Lessons 2.4, 2.6, 2.7	Exercises 13, 15, 16	Review 14, 15 Mixed Practice 18
Subtracting whole numbers	Lessons 3.3, 3.5–3.7	Exercises 20, 22–24	Review 24 Practice 27, 28
Multiplying whole numbers	Lessons 4.2, 4.4, 4.5, 4.8	Exercises 29, 30, 33	Review 36, 37 Practice 40, 42
Recognizing clue words for addition	Lesson 2.5	Exercise 14	Practice 16
Recognizing clue words for subtraction	Lesson 3.4	Exercise 21	Practice 26
Recognizing clue words for multiplication	Lesson 4.6	Exercise 31	Practice 41

Diagnostic and Placement Guide

Chapter 5 Dividing Whole Numbers

The Percent Accuracy scores are based on the number of problems in a lesson that have been answered correctly.

Resource Planner

After diagnosing your students' needs, use the correlating Program Resources for reinforcement, reteaching, or enrichment. Additional activities for customizing the lessons can be found in this guide.

KEY
Reteaching = ⤴   Reinforcement = ↓   Enrichment = ⤴

	Lessons	Percent Accuracy				Program Resources			
		50%	65%	80%	100%	↓ Student Edition • Extra Practice	↓ Workbook • Exercises	Teacher's Planning Guide	Classroom Resource Binder
5.1	What Is Division?	4	5	6	8		35	↓ p. 49	
5.2	Basic Division	8	10	12	15	p. 419	36	⤴ p. 49	
5.3	Dividing with Remainders	18	23	29	36	p. 419	37	↓ p. 50	↓ Review 50 Dividing with Remainders ↓ Practice 54 Dividing with Remainders
5.4	Dividing Larger Numbers	12	16	19	24	p. 419 p. 420	38	↓ p. 51	↓ Review 51 Dividing Larger Numbers ↓ Practice 56 Dividing by Numbers with More Than One Digit
5.5	Checking Division	4	5	6	8	p. 419 p. 420	39	⤴ p. 51	
5.6	Problem Solving: Clue Words for Division	1	2	2	3	p. 420	40	⤴ p. 52	↓ Practice 55 Problem Solving: Clue Words for Division
5.7	Dividing by Numbers with More Than One Digit	12	16	19	24	p. 420	41	⤴ p. 53	↓ Review 52 Dividing by Numbers with More Than One Digit ↓ Practice 56 Dividing by Numbers with More Than One Digit ↓ Mixed Practice 59 Dividing by Numbers with One or More Digits ⤴ Challenge 62 Who Is It?
5.8	Zeros in the Quotient	13	17	21	26	p. 420	42	↓ p. 54	↓ Review 53 Zeros in the Quotient ↓ Practice 57 Zeros in the Quotient ↓ Mixed Practice 60 Addition, Subtraction, Multiplication, and Division
5.9	Problem Solving: Choose the Operation	2	3	4	5	p. 420	43	⤴ p. 54	↓ Practice 58 Problem Solving: Choose the Operation

Chapter 5

Lessons	Percent Accuracy				Program Resources			
	50%	65%	80%	100%	Student Edition Extra Practice	Workbook Exercises	Teacher's Planning Guide	Classroom Resource Binder
Math in Your Life: Determining Miles per Gallon (mpg)	See page 55 in this guide for assessment objectives.						p. 55	Math in Your Life 61 Determining Miles per Gallon (mpg)
5.10 Estimating and Thinking	1	2	2	3		44	p. 56	
Chapter 5 Review								
Vocabulary Review	5	6.5 (writing is worth 4 points)	8	9			p. 57	Vocabulary 49
Chapter Quiz	13	16	20	25			p. 58	Chapter Test A, B 63–64

Customizing the Chapter

Opening the Chapter
Student Edition pages 84–85

Photo Activity
Procedure: Discuss rates at which various events take place. Have students find the rate of a car that travels 80 miles in 2 hours. Ask them how they would find how many miles that car would travel in 1 minute. Guide them to see that rates are found by dividing by time. Have students use reference materials to find examples of rates that occur yearly, monthly, daily, or hourly. For each one, have them discuss how they would find the rate for smaller units of time.

Example: If 2,400 trees are cut down every 8 hours, how many trees are cut down every hour? every minute?

Answers: 2,400 ÷ 8 = 300 per hour; 300 ÷ 60 = 5 per minute.

Words to Know
Review with the students the Words to Know on page 85 of the Student Edition. Help students remember these terms by having them brainstorm ways to distinguish among the words *dividend*, *divisor*, and *quotient*.

The following words and definitions will be covered in this chapter.

division the process of finding how many times one number contains another.

dividend the number to be divided

divisor the number to divide by

quotient the number obtained by dividing one number into another; the answer in a division problem

remainder the number left over in a division problem

Life Journal Project
Summary: Students look for examples of division in daily life and write problems based on these examples.

Materials: paper, pencils

Procedure: Provide students with examples. Point out that division implies separating a set into *equal amounts*. Students can complete this project in class, or they can keep a journal throughout the chapter and cite examples found during their daily routine at home and at school.

Assessment: Use the Individual Activity Rubric on page *xi* of this guide. Fill in the rubric with the additional information below. For this project, students should have:

• made a list of ways division is used everyday.
• written a word problem for each item on their list.

Learning Objectives
Review the Learning Objectives on page 85 of the Student Edition before starting the chapter. Students can use the list as a learning guide. Suggest they write the objectives in a journal or use CM6 *Goals and Self-Check* in the Classroom Resource Binder.

After each lesson, have students write an example of the skill they learned under the appropriate objective. Suggest that students use these notes as a learning guide to help them study for the chapter test.

5.1 What Is Division?
Student Edition page 86

Prerequisite Skill
- Making equal groups

Lesson Objective
- Write division problems in horizontal and computational form using numbers.

Words to Know
division, dividend, divisor, quotient

Cooperative Group Activity

12-Pack Puzzle
Materials: counters, paper, pencils

Procedure: Organize students into groups of four. Give each group 12 counters. Write the following on the chalkboard: *Luke bought a 12-pack of soda to bring to his after-school club. He wanted to find out how many cans each person could have if the following numbers of students showed up at the meeting: 2, 3, 4, 6, and 12.*

- Students use the counters to represent the 12 cans of soda. The group determines how many cans of soda each student in the word problem would get if only two students showed up at the meeting.
- One student uses the counters to model the situation. In their Math Journal, students draw a picture of the counters and write the division in two ways.
- Repeat the procedure for the other numbers of students who show up at the meeting.

Customizing the Activity for Individual Needs
ESL Help students understand *divide* by showing the word, symbols, and an illustration of 6 ÷ 2. Have students explain in their own words what division means, either directly or through a real-life example.

Learning Styles Students can:

 draw pictures to model division problems.

 write the numbers in a division problem on self-stick notes and practice going from one form to the other.

 explain how to change a division problem from one form to the other.

Reinforcement Activity

Have student pairs draw pictures representing a division problem, such as the one shown below. Partners exchange papers and write the division problem in both forms.

$$8 \div 2 = 4 \qquad 2\overline{)8}^{4}$$

Alternative Assessment

Students can write a division problem based on a picture of objects divided into groups.

Example: Write a division problem suggested by this picture in two forms.

○○○○ ○○○○

Answer: $8 \div 4 = 2$ and $4\overline{)8}^{2}$

5.2 Basic Division
Student Edition page 87

Prerequisite Skills
- Understanding division
- Reading tables

Lesson Objective
- Use a multiplication table for basic division facts.

Cooperative Group Activity

What's the Problem?
Materials: multiplication table, colored pencils, index cards

Procedure: Organize students into groups of three. Provide one copy of a multiplication table for each group.

- One student colors in a random number in the table. The other students locate the numbers at the beginning of the row and the top of the column that contains the chosen number.

- Group members work together to write a division problem using the three numbers. Students can then write the division problem on one side of an index card and the quotient on the other side.
- Students repeat the activity, taking turns choosing the number in the table.

Customizing the Activity for Individual Needs

ESL Provide students with practice using the words *row* and *column*. Point to a number in the table and ask students: "What is the number at the beginning of this row? What number is at the beginning of this column?"

Learning Styles Students can:

 use colored markers to highlight the paths from the divisor to the dividend to the quotient to find division facts in the multiplication table.

 use counters to find the basic division facts. They then use the multiplication table to verify the results.

 explain to a partner how the table can be used to find quotients, such as 42 ÷ 6.

Enrichment Activity

Provide students with counters and a multiplication table. Have them sort the counters into equal groups. Then have them write related multiplication and division facts, using the table as a reference.

Alternative Assessment

Students can demonstrate mastery of basic division facts by responding orally to flash cards.

Example: Say the quotient shown on the card.

Answer: 6

5.3 Dividing with Remainders
Student Edition pages 88–89

Prerequisite Skills
- Understanding place value
- Knowing basic multiplication facts
- Adding larger numbers

Lesson Objectives
- Find the remainder in division problems.
- Life Skill: Make seating arrangements.

Words to Know
remainder

Cooperative Group Activity

Any Leftovers?

Materials: counters, paper, pencils

Procedure: Organize students into groups of three. Give each group 25 counters. Write a number on the chalkboard equal to or less than 25.
- Students use only that number of counters and divide them into 3 equal groups.
- Group members write the division problem modeled by their counters, including the remainder. If the remainder is zero, they do not have to write it.
- Repeat the activity for other dividends and divisors.

Customizing the Activity for Individual Needs

ESL Help students better understand the word *remainder*. Have them use the words *remainder* and *remaining* in non-math-related phrases to indicate "the part that is left."

Examples: "the remainder of the day"; "the remaining pizza."

Learning Styles Students can:

 highlight the remainder that occurs in the division process and the remainder with the quotient.

 use counters to model division problems.

 explain to a partner what the remainder means.

Reinforcement Activity

Have students use a multiplication table to write the basic fact closest to a given division problem. For example, for 16 ÷ 5, students need to find out how many times 5 will go into 16 without going over 16. They follow along the 5 row and see that 15 is as high as they can go without going over 16. Students write the fact 15 ÷ 5 = 3.

Alternative Assessment

Students can write a division problem with a remainder based on a model.

Example: What division problem is being modeled?

Answer: 13 ÷ 3 = 4 R1

5.4 Dividing Larger Numbers
Student Edition pages 90–91

Prerequisite Skills
- Knowing basic multiplication facts
- Knowing basic division facts

Lesson Objectives
- Divide 2- and 3-digit numbers by 1-digit numbers.
- Life Skill: Divide food into equal amounts.

Cooperative Group Activity

Roll and Divide

Materials: number cubes, paper, pencils

Procedure: Organize students into groups of four. Give each group four number cubes.
- Each student rolls a number cube. Students form one 3-digit number to use as the dividend and use the remaining number as the divisor.
- Each student does the division problem individually.
- Students compare their results. They look at each other's work to find any errors if the answers don't agree.
- Students repeat the activity several times.

Customizing the Activity for Individual Needs
ESL Help students understand that *bring down* means to rewrite a digit in the dividend next to the difference. Emphasize these words while modeling a problem on the chalkboard.

Learning Styles Students can:

 write division problems on grid paper to help keep digits aligned.

 use digit cards to model longer division problems, lining up the cards in the quotient with their proper place values.

 solve division problems by thinking out loud, explaining each step.

Reinforcement Activity
Tell students you want to read a book in 7 days. The book has 427 pages. Have them find the number of pages you have to read each day to finish. Repeat for books with 317 pages and 192 pages. Discuss the significance of any remainders.

Answers: 427: 61; 317: 45 R2; 192: 27 R3

Alternative Assessment
Students can identify and explain where to place the first digit when dividing larger numbers.

Example: Where do you place the first digit in the quotient for 236 ÷ 7?

Answer: Since 7 does not divide into 2, the first digit in the quotient will be in the tens place.

5.5 Checking Division
Student Edition page 92

Prerequisite Skills
- Multiplying with one regrouping
- Multiplying larger numbers

Lesson Objective
- Check answers to division problems.

Cooperative Group Activity

Check It Out

Materials: paper, pencils, number cubes

Procedure: Organize students into groups of three. Give each group four number cubes.
- One student rolls the number cubes to form a 1-digit number for the divisor and a 3-digit number for the dividend.
- Another student does the division.
- The third student checks the division.
- Students repeat the activity, rotating roles.

Customizing the Activity for Individual Needs
ESL To help students differentiate among *divisor, dividend, quotient,* and *remainder,* have pairs solve the problem 641 ÷ 8. Then take turns labeling parts with the proper words. They then can label the words in the check.

Learning Styles Students can:

 write color-coded division problems in a notebook. Then use the same color code to write the check for each problem.

 write the parts of a division problem on separate index cards, then rearrange the cards to write the check.

 explain how to check a division problem aloud.

Reteaching Activity

Write a division problem on the chalkboard. Have students write the divisor, dividend, quotient, and remainder on separate self-stick notes. Have them position the notes to match the problem on the chalkboard. Then rearrange the notes to show the multiplication that would be used to check the answer.

Alternative Assessment

Students can check division using color coding for quotient, dividend, divisor, and remainder.

Example: Use different colors to code the quotient, dividend, divisor, and remainder in this division problem. Use the same color code to check the answer.

264 ÷ 5 = 52 R4

Answer: Students should write the example below, using the same colors for the numbers as they used in the original problem.

```
   52
 × 5
  260
 + 4
  264
```

USING YOUR CALCULATOR
Checking Division
Student Edition page 93

Lesson Objective
• Life Skill: Use a calculator to check division, addition, subtraction, and multiplication.

Activity
Match and Check
Materials: calculators, slips of paper, pencils

Procedure: Organize students into pairs. Provide each pair with 8 slips of paper. Have them write the following problems on four separate slips: 84 ÷ 6, 97 − 74, 4 × 21, 27 + 29. Have them write these answers on the remaining four slips: 23, 56, 84, 14.

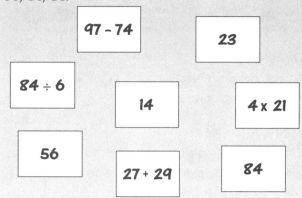

- Students try to match the answers with the problems, without calculating. Then they take turns checking the answers on the calculator. Each student scores a point for each correct match.
- Next, students make up four problems of their own, using 1- and 2-digit numbers. They write their problems and answers on separate slips of paper and exchange slips.
- Students match problems and answers using mental math, then use the calculator to check, scoring a point for each correct match.

5.6 Problem Solving: Clue Words for Division
Student Edition pages 94–95

Prerequisite Skill
• Dividing by one-digit divisors

Lesson Objectives
• Recognize clue words for division.
• Solve word problems using division.
• Life Skill: Use a formula to find rate or time.

Cooperative Group Activity

Clues Cards

Materials: index cards, paper, pencils

Procedure: Write the following clue words on the chalkboard: *how much did each, how many times, into how many.* Organize students into pairs.

Distribute index cards to each group. Assign each pair of students different clue words to use. Tell students to use 1-digit divisors in their problems.

- Each pair writes a word problem on the front of the index card using the assigned clue words, then writes the solution on the back of the card.
- When all pairs are finished writing their problems, a volunteer collects the cards. The student places them problem-side up in a pile in the front of the room.
- Pairs pick a card from the pile without looking. They solve the problem together by identifying the clue words, then showing their work.
- Students check their answers by looking at the back of the index cards.

Customizing the Activity for Individual Needs

ESL Help students understand the difference between the phrases *how much* and *how many*. Explain that *much* means *amount of* and *many* means *number of*. Make a chart of the phrases and discuss which of the following words go under each heading: time, books, hours, rice, butter, bananas.

Answers: how much: time, rice, butter; how many: books, hours, bananas

Learning Styles Students can:

 highlight division clue words in word problems.

 write clue words on separate index cards, then match a problem to one of the cards.

 read problems from their books, stressing the division clue words in a louder voice.

Enrichment Activity

Have students write two problems that can be solved using these formulas:

$$\text{Time} = \frac{\text{Distance}}{\text{Rate}} \qquad \text{Rate} = \frac{\text{Distance}}{\text{Time}}$$

Students then trade problems and solve.

Alternative Assessment

Students can identify clue words in a division problem, then explain how to solve the problem.

Example: What clue words in this problem tell you to divide? How would you solve the problem?

> The drama club sold all 822 tickets for the 3 nights of their show. They sold the same number of tickets of each night. How many tickets did they sell for each night?

Answer: Accept all reasonable explanations. Clue words: *how many . . . for each*; 274

5.7 Dividing by Numbers with More Than One Digit
Student Edition pages 96–97

Prerequisite Skill
- Dividing by 1-digit divisors

Lesson Objectives
- Divide 3-digit numbers by 2-digit numbers.
- Life Skill: Read distances on a map.

Cooperative Group Activity

Step-by-Step

Materials: index cards, paper, pencils

Procedure: Write the division problems 486 ÷ 21 and 837 ÷ 32 on the chalkboard. Organize students into groups of four. Give each group four index cards. Instruct each group to write the words *divide*, *multiply*, *subtract*, and *bring down* on separate index cards. Have each member choose one of the four cards.

- Students copy the first division problem on a sheet of paper. The group does the division. Each member does the step written on his or her card, repeating the steps as needed.
- Members check their answers by multiplying and adding any remainders.
- Group members trade cards and repeat the activity for the second problem.

Answers: 23 R3; 26 R5

Customizing the Activity for Individual Needs

ESL Help students understand that an *average* speed of 55 miles per hour means that they can think of the speed as 55 miles per hour for the entire trip. The word *average* implies division.

Learning Styles Students can:

 color code the steps in an example with the words that describe the process.

 write names and symbols on index cards, then place the cards in the order they would appear as they do the division.

 say the steps they are writing when it is their turn.

Enrichment Activity

Provide students with local maps. Have them choose two cities and find the distance between them. Ask students to determine how long it would take to drive between the cities at different rates. Repeat for two different cities.

Alternative Assessment

Students can determine where to place the first digit in a quotient when dividing by a 2-digit number.

Example: For each division below tell where you would place the first digit in the quotient.

$92\overline{)828}$ $22\overline{)594}$

Answers: ones place; tens place

5.8 Zeros in the Quotient
Student Edition pages 98–99

Prerequisite Skill
- Dividing by 1- and 2-digit numbers

Lesson Objectives
- Divide when there is a zero in the quotient.
- Life Skill: Calculate installment plan payments.

Cooperative Group Activity

The Purpose of Zero

Materials: paper, pencils

Procedure: Write $2\overline{)600}$, $2\overline{)608}$, and $2\overline{)680}$ on the chalkboard. To the right of the problems write *3, 30, 34, 300, 304, 340.*

- As someone points to $2\overline{)600}$, a volunteer picks a number, either *3, 30, 34, 300, 304,* or *340,* and tells why the number could be the quotient.
- A volunteer does the division on the chalkboard, emphasizing that without zero as a placeholder, the quotients 3 and 230 would be too small. (The correct answer is 300.)
- Repeat until all the division problems have been matched to a quotient.
- To extend the activity, small groups match $3\overline{)1200}$, $3\overline{)1209}$, and $3\overline{)1290}$ with these possible quotients: 40, 43, 400, 403, 430, and 4,000. Each group justifies its choices.

Customizing the Activity for Individual Needs

ESL Use role playing to demonstrate the idea of *placeholder.* Have several students stand in line, as though they were waiting to buy lunch. Have one student pretend that he or she has to leave his or her place in line for a moment. Have the student ask another student to hold his or her place in line. Relate this concept to that of needing to "hold a place" in a quotient by writing a zero.

Learning Styles Students can:

 write placeholder zeros in a different color.

 do the division at the chalkboard.

 read aloud the number that needs to be brought down and the number that is put in the quotient.

Reinforcement Activity

Suggest that students plan to buy a $176 CD player on an installment plan. They plan to pay $50 for the down payment and $9 each month. Ask them to find how many months it will take before it is paid off.

Answer: 14 months

Alternative Assessment

Students can explain why a division problem is incorrect.

Example: Explain why this division problem is incorrect. Show the correct division.

1,825 ÷ 6 = 34 R1

Answer: A zero is needed in the tens place of the quotient. The correct quotient is 304 R1.

```
    304 R1
6)1,825
   -18
    025
    -24
      1
```

5.9 Problem Solving: Choose the Operation
Student Edition pages 100–101

Prerequisite Skill
- Knowing clue words for addition, subtraction, multiplication, and division

Lesson Objective
- Identify and use the appropriate operation to solve a word problem.

Cooperative Group Activity

Class Creations

Materials: pencils, paper

Procedure: Organize students into groups of four or five. Write the following as prices from a store in the mall: 5 pens for $10, poster $15, picture frame $20, calendar $12, 3 magnets for $9.

- Groups write four word problems, one for each operation, using the numbers and items listed on the chalkboard.
- Each group also writes a solution key for the four problems on a separate sheet of paper.
- Groups exchange problems and solve.
- Groups then return the solutions to be checked by the writers of the word problems.
- Students can then *go shopping* and determine what they can buy for $35. Have them come as close to $35 as they can without going over.

Possible answer: a picture frame, a calendar, and one magnet.

Customizing the Activity for Individual Needs

ESL To help students remember the clue words for the various operations, write the words and the symbols for each operation on separate index cards.

CLUE WORDS			
Add	Subtract	Multiply	Divide
in all	how many more	total	how many times
total	difference	of	into how many
altogether	how many fewer (less)	at	how much did each
		for	
		in each	

Mix up the cards. Have the students place the clue words in the pile with the correct symbol.

Learning Styles Students can:

 draw a picture of the word problem to help them decide on the operation.

 use play money to model solutions to the word problem.

 explain why they chose a particular operation to solve a problem.

Reteaching Activity

Have a student make a tape recording of word problems such as those on page 101 of the Student Edition. The student should point out clue words and explain the plan to solve each problem. After checking the recording for clarity, instruct students who need further help to replay the tape while reading along with the text.

Alternative Assessment

Students can determine which operation to use to solve problems they've written.

Example: Choose two different operations. Then use the following information to write two word problems that can be solved with the operations you chose. Underline the clue words, and solve your problems.

A pizza costs $7.20 and has 8 slices.

Sample answers: What is the cost of each slice of pizza?

$7.20 ÷ 8 = $.90.

What would be the total cost of 4 pizzas?

$7.20 × 4 = $28.80.

MATH IN YOUR LIFE
Determining Miles per Gallon (mpg)   *Student Edition page 102*

Chapter 5

Lesson Objectives

- Apply division to find miles per gallon.
- Life Skill: Compare car features.
- Communication Skill: Write a letter requesting car brochures.
- Communication Skill: Negotiate car prices.

Activities

Look It Up

Materials: car brochures or newspaper ads, *Consumer Reports*, Internet car sources

Procedure: Have students look up and write descriptions of different vehicle classes. For example, students can research compact cars, luxury vehicles, and another class of vehicle of their choice. Suggest that students include how each type of vehicle is generally used.

Write a Business Letter

Materials: phone book, paper and pen or computer word processor

Procedure: Have students write letters to several car dealers. They can find addresses in the Yellow Pages of the local phone book. Have them

(Continued on p. 56)

(Continued from p. 55)
request brochures and other documents that will provide the following information:
- which models are most popular for different age groups
- the selling features of each model
- average miles per gallon for each model
- available finance plans

Act It Out
Materials: Consumer Reports

Procedure: Have two students role-play buying a car. One student is the consumer, the other student is the car dealer. The consumer needs to:
- check *Consumer Reports* for the safety and repair record of the car
- call other dealerships for price quotes on the same car

The dealer needs to be familiar with the car's features, including:
- safety features
- features that enhance the car's performance
- gas mileage

With this information, have the consumer negotiate with the dealer for the best possible price for the car.

Practice
Have students complete Math in Your Life 61 *Determining Miles per Gallon (mpg)* from the Classroom Resource Binder.

Internet Connection
Students can use the following Web site to research and compare features of different vehicles:
  AutoVantage
    http://www.autovantage.com

They can find links to other vehicle purchasing sites and information by choosing the Automotive link from:
  The HotBot Shopping Directory Channel
    http://www.hotbot.com/shop

5.10 Estimating and Thinking
Student Edition page 103

Prerequisite Skill
- Dividing whole numbers

Lesson Objective
- Use estimation to choose the best answer.

Cooperative Group Activity
Matching Up
Materials: flyers from department stores, pharmacies, or grocery stores; index cards, pencils, paper

Procedure: Provide flyers or have students bring them in. Organize students into groups of three. Have each group choose a flyer. Give each group six index cards. Have the groups make the prices a whole number's price by crossing out the "cents."

- Each group makes up three different word problems based on items and prices in the flyer. They write each problem on separate index cards.
- Students solve their problems, making sure that each problem has a different answer. They write the answers on separate index cards.
- Groups exchange index cards. Group members use reasoning and estimation to match the answers with the problems.

Customizing the Activity for Individual Needs
ESL Help students understand the word *estimate*. Brainstorm other words that can imply estimating, such as *about, approximately, best guess, on average*. Have students use each word in a sentence.

Learning Styles Students can:

 use different colors to cross out answer choices that don't make sense: one color for answers that are too big; another color for answers that are too small.

 place counters over answers that they rule out as being too large or too small.

 listen as another student "thinks aloud" to find the most reasonable answer.

Enrichment Activity

Have students review how to round whole numbers (see Lesson 1.9). Have them estimate a solution to the following problem:

> The Greenville Road Race raised $1,800 for charity. There were 33 runners in the race. About how much money did each runner raise?

Answer: 1,800 ÷ 30 = 60; about 60

Alternative Assessment

Students can choose the best estimate by analyzing each choice.

Example: Analyze each choice for the following problem, then choose the best answer.

> About how many miles per gallon will Karen's car get if she drives 112 miles on 5 gallons?
> (a) 2   (b) 120   (c) 20

Answer: Accept all reasonable explanations. Students might say (a) is too small, (b) is too big, (c) is a good estimate because 100 ÷ 5 = 20.

Closing the Chapter
Student Edition pages 104–105

Chapter Vocabulary

Review with the students the Words to Know on page 85 of the Student Edition. Then have students quiz each other in pairs.

Have students copy and complete the Vocabulary Review questions on page 104 of the student edition.

For more vocabulary practice, have them complete Vocabulary 49 *Dividing Whole Numbers* from the Classroom Resource Binder.

Test Tips

In pairs, students can take turns showing how to solve a problem by using one of the test tips.

Learning Objectives

Have the students review CM6 *Goals and Self-Check*. They can check off the goals they have reached. Note that each section of the quiz corresponds to a Learning Objective.

Group Activity

Summary: Students record examples of unit pricing and create word problems about finding the unit price of an item. Encourage students to use whole numbers.

Materials: dictionary, pencils, paper

Procedure: You may wish to have group members find different types of items, such as produce, meat, cleaning products, or canned goods. If students cannot visit a supermarket, provide supermarket circulars or flyers for reference. Encourage students to be creative with their word problem presentations. They might cut out examples of the items in the problem and make a poster for a bulletin board display. Challenge groups to solve or add onto other groups' problems.

Assessment: Use the Group Activity Scoring Rubric on page *xii* of this guide. Fill in the rubric with the additional information below. For this project, students should have:

- written the meaning of *unit price* and created an appropriate word problem.
- recorded ten examples of unit pricing.

RELATED MATERIALS See the unit overview page for other Globe Fearon books that can be used to enrich and extend the material in this unit.

Assessing the Chapter

Traditional Assessment

Chapter Quiz
The Chapter Quiz on pages 104–105 of the Student Edition can be used as either an open- or closed-book test, or as homework. Use the quiz to identify concepts in the chapter that students need to review and practice.

Chapter Tests
Use Chapter Test A Exercise 63 and Chapter Test B Exercise 64 *Dividing Whole Numbers* from the Classroom Resource Binder to further assess mastery of chapter concepts.

Additional Resources
Use the Resource Planner on page 47 to assign additional exercises from the Classroom Resource Binder and Workbook.

Alternative Assessment

Interview
Write the problem $3{,}147 \div 65$ on a piece of paper. Ask:
- Where should the first digit in the quotient be placed?
- How can you find the first digit?
- What steps do you follow after you place the first digit in the quotient?
- What is the quotient?
- How could you check your division?

Application
Write a word problem on the chalkboard.

> There are 4,287 sheets of paper for 21 classes. Each class is given the same amount of paper. How many sheets of paper will each class be given?

Have students:
- decide on a reasonable answer to the problem.
- solve the problem.
- explain why zero as a placeholder was needed.
- explain the significance of the remainder in relation to the word problem.

Planning the Chapter

Chapter 6 • More About Numbers

Chapter at a Glance

SE page	Lesson	
106		Chapter Opener and Project
108	6.1	Divisibility Tests for 2, 5, and 10
110	6.2	Divisibility Tests for 3, 6, and 9
112	6.3	Divisibility Test for 4
113		Using Your Calculator: Test for Divisibility
114	6.4	Factors and Greatest Common Factor
116	6.5	Multiples and Least Common Multiple
118	6.6	Prime Numbers
119	6.7	Exponents
120	6.8	Squares and Square Roots
121		On-The-Job Math: Electric Meter Reader
122	6.9	Problem Solving: Extra Information
124		Chapter Review and Group Activity
126		Unit Review

Learning Objectives
- Use divisibility tests.
- Find the factors of a number.
- Find the multiples of a number.
- Write prime factorizations.
- Find squares and square roots.
- Solve problems with extra information.
- Apply number sense to reading electrical meters.

Life Skills
- Make arrangements.
- Make ordering decisions.
- Use a calculator to test for divisibility.
- Make a seating plan.
- Coordinate appointment dates.
- Calculate the cost of using electricity.
- Solve different problems based on the same situation.

Communication Skills
- Write a letter.
- Explain how to estimate electricity usage.
- Read an electric bill.

Workplace Skills
- Read and record electricity use on a meter.
- Estimate a customer's electricity usage.

PREREQUISITE SKILLS
To assess mastery of prerequisite skills, use a selection of exercises from any of the program resources referenced below. The same resources can be used to provide remediation if necessary.

	Program Resources for Review		
Skills	Student Edition	Workbook	Classroom Resource Binder
Using basic addition facts	Lesson 2.2	Exercise 11	
Using basic multiplication facts	Lesson 4.2	Exercise 27	
Using basic division facts	Lesson 5.2	Exercise 36	
Dividing with remainders	Lesson 5.3	Exercise 37	Review 50 Practice 54
Multiplying with one regrouping	Lesson 4.4	Exercise 29	Review 36 Practice 39

Diagnostic and Placement Guide

Chapter 6 More About Numbers

The Percent Accuracy scores are based on the number of problems in a lesson that have been answered correctly.

Resource Planner

After diagnosing your students' needs, use the correlating Program Resources for reinforcement, reteaching, or enrichment. Additional activities for customizing the lessons can be found in this guide.

KEY
Reteaching = ⤺   Reinforcement = ⬇   Enrichment = ⤻

Lessons		Percent Accuracy				Program Resources			
		50%	65%	80%	100%	Student Edition Extra Practice	Workbook Exercises	Teacher's Planning Guide	Classroom Resource Binder
6.1	Divisibility Tests for 2, 5, and 10	10	13	16	20	p. 421	45	⬇ p. 61	Visual 4 *Whole Numbers to 100*
6.2	Divisibility Tests for 3, 6, and 9	13	16	20	25	p. 421	46	⤺ p. 62	Visual 4 *Whole Numbers to 100*
6.3	Divisibility Test for 4	12	16	19	24	p. 421	47	⬇ p. 63	Visual 4 *Whole Numbers to 100*
6.4	Factors and Greatest Common Factor	10	13	16	20	p. 421	48	⬇ p. 64	⬇ Review 66 *Greatest Common Factor* ⬇ Practice 70 *Factors and Greatest Common Factor*
6.5	Multiples and Least Common Multiple	13	17	21	26	p. 421	49	⤺ p. 64	⬇ Review 67 *Multiples and Least Common Multiple* ⬇ Practice 71 *Multiples and Least Common Multiple* Visual 4 *Whole Numbers to 100*
6.6	Prime Numbers	5	7	8	10	p. 421	50	⤺ p. 65	⬇ Review 68 *Prime Numbers* Visual 4 *Whole Numbers to 100*
6.7	Exponents	8	10	12	15	p. 421	51	⬇ p. 66	
6.8	Squares and Square Roots	7	8	10	13	p. 421	52	⬇ p. 67	⬇ Review 69 *Squares and Square Roots*
	On-The-Job Math: Electric Meter Reader	*Can be used for portfolio assessment.*						⤺ p. 68	⤺ On-The-Job Math 73 *Electric Meter Reader*
6.9	Problem Solving: Extra Information	1	2	2	3		53	⬇ p. 68	⬇ Practice 72 *Problem Solving: Extra Information* ⤺ Challenge 74 *Using Stamps*
Chapter 6 Review									
	Vocabulary Review	7	9	11	14			⬇ p. 69	⬇ Vocabulary 65
		(writing worth 4 points)							
	Chapter Quiz	10	13	16	20			⬇ p. 70	⬇ Chapter Test A, B 75, 76
Unit 1 Review		5	6.5	8	10			⬇ p. 70	⬇ Unit Test T7
		(critical thinking worth 4 points)							

Customizing the Chapter

Opening the Chapter
Student Edition pages 106–107

Photo Activity
Teacher Background: Professional photographers use a method called bracketing to compensate for uncertainties in exposure. For a single subject or scene, they may take a series of exposures progressively varying adjustments in aperture, speed, or F-stop settings.

Procedure: Have students share experiences taking photographs. Ask: "Did you use film with 24 or 36 exposures on a roll? What did you photograph? Suppose that for each person, object, or scene you wanted to photograph, you needed to take four shots to be sure of a good one. How many "different scenes" could you photograph from a roll with 24 exposures? from a roll with 36 exposures?"

Answers: 6 scenes with 24 exposures; 9 scenes with 36 exposures

Words to Know
Review with the students the Words to Know on page 107 of the Student Edition. Help students remember these words by having them identify related pairs of words, such as factors/multiples; prime number/composite number; square/square root. The following words and definitions are covered in this chapter.

divisible can be divided without a remainder

factors numbers multiplied to get a product

greatest common factor (GCF) the largest factor that two or more numbers share

multiples possible products of a given number

least common multiple (LCM) the smallest multiple that two or more numbers share

prime number number with only itself and 1 as factors

composite number number with more than two factors

exponent tells how many times to use a number as a factor

square product of multiplying a number by itself

square root number that was squared

Class Team Project
Summary: Students use the number of students in each of their classes to list all the possible equal-sized teams that could be made for each class.

Materials: pencils, paper

Procedure: Students can complete the first part of this project in one school day by counting and writing down the number of students in each of their classes. They may complete the calculations during math class or as homework after completing Lesson 6.3.

Assessment: Use the Individual Activity Rubric on page *xi* of this guide. Fill in the rubric with the additional information below. For this project, students should have:
- found the number of students in at least three different classes.
- found all the factors of these three numbers.

Learning Objectives
Review with the students the Learning Objectives on page 107 of the Student Edition before starting the chapter. Students can use the list of objectives as a learning guide. Suggest that they write the objectives in a journal or use CM6 *Goals and Self-Check*.

After each lesson, have students write an example of the skill they learned under the appropriate objective. Suggest that students use these notes as a learning guide to help them study for the chapter test.

6.1 Divisibility Tests for 2, 5, and 10 *Student Edition pages 108–109*

Lesson Objectives
- Use divisibility tests for 2, 5, and 10.
- Life Skill: Use a floor plan to assign camping cabins.

Words to Know
divisible

Cooperative Group Activity

Charting Circles
Materials: table of numbers 1 to 100 or Visual 4 *Whole Numbers to 100*; tape; a green, red, and blue marker

Procedure: Make three copies of Visual 4 and one enlarged copy. Tape the enlarged copy to the chalkboard. Organize students into three groups. Give each group a copy of Visual 4 and a different-colored marker. Assign each group a number: 2, 5, or 10.

- Each group finds all the numbers in the table that are divisible by the assigned number. A student in each group circles them with the marker.
- A representative from each group writes the circled numbers on the chalkboard.
- Volunteers point out numbers that are circled more than once and explain what that means.

Customizing the Activity for Individual Needs

ESL To increase understanding of the word *divisible*, illustrate *divisible* and *not divisible*. Draw 12 boxes on the board. Say, "divisible": circle groups of 6 (none left over). Say, "not divisible": circle groups of 5 with 2 boxes left over. Let volunteers try the activity with 24 boxes drawn on the chalkboard.

Learning Styles Students can:

 use different colors to identify numbers divisible by 2, 5, and 10 on the same number table.

 use counters to mark numbers in a table.

 say the divisibility rule aloud as a question, using any number to test it. *Example:* Does 20 end in a 0 or 5? Yes, so it is divisible by 5.

Reteaching Activity

Provide 20 counters for a student to divide into groups of 2. Ask: "Are there any counters left over? Is 20 divisible by 2?" Repeat with groups of 5 and groups of 10. Try this with nonfactors, too, such as 3, 6, or 9. Explain that leftovers mean that it is not divisible.

Alternative Assessment

Students can determine if a number is divisible by 2, 5, or 10 and explain their reasoning.

Example: Is 15 divisible by 5? How do you know? Is 15 divisible by 10? How do you know?

Answer: Accept all reasonable explanations. Students may say that 15 is divisible by 5 because the last digit is 5; it is not divisible by 10 because the last digit is not 0.

6.2 Divisibility Tests for 3, 6, and 9 *Student Edition pages 110–111*

Prerequisite Skills
- Using the divisibility test for 2
- Using basic multiplication facts
- Using basic division facts

Lesson Objectives
- Use divisibility tests for 3, 6, and 9.
- Life Skill: Decide the amount of materials needed for a given situation.

Cooperative Group Activity

Charting Number Patterns

Materials: table of numbers 1 to 100 or Visual 4 *Whole Numbers to 100*; tape; a green, red, and blue marker

Procedure: Make three copies of Visual 4 and one enlarged copy. Tape the enlarged copy to the chalkboard. Organize students into three groups. Give each group a copy of Visual 4 and a different-colored marker. Assign each group a number: 3, 6, or 9.

- Each group finds all the numbers on the table that are divisible by the assigned number. A student from each group circles them with the marker.
- A representative from each group writes the circled numbers on the chalkboard.
- Volunteers point out numbers that are circled more than once and explain what that means.

Customizing the Activity for Individual Needs

ESL Review the words *divisible* and *sum* by brainstorming definitive phrases with the class. *Example: Divisible* means "goes into evenly" or "divides into with none left over." *Sum* means *total, to add,* or *altogether*. Write a division and an addition example on the chalkboard. Point to each one and ask a volunteer to decide whether the word is *divisible* or *sum*.

Learning Styles Students can:

 use the colored markers to identify number patterns, such as all numbers divisible by 9 are also divisible by 3.

 use colored counters to mark numbers divisible by 3, 6, and 9 on the table.

 describe the steps they use to test for divisibility by their assigned number.

Enrichment Activity

Give each student a spinner divided into three equal sections labeled 3, 6, and 9. Have students spin the spinner, then write a list of at least seven multidigit numbers that are divisible by the number shown on the spinner. Students then check their numbers by applying the divisibility tests.

Alternative Assessment

Students can determine if a number is divisible by 3, 6, or 9 and explain their reasoning.

Example: Is 18 divisible by 6? How do you know? Is 18 divisible by 9? How do you know?

Answer: A reasonable explanation might be "18 is divisible by 6 because it is even and 1 plus 8 is 9, which is divisible by 3. It is divisible by 9 because 9 (1 + 8) divided by 9 is 1."

6.3 Divisibility Test for 4
Student Edition page 112

Prerequisite Skill
- Using basic multiplication and division facts

Lesson Objective
- Use the divisibility test for 4.

Cooperative Group Activity

Charting Fours
Materials: a 10 × 10 grid of blank squares or Visual 12 *Hundredths Model*, pencils, markers, tape

Procedure: Organize students into small groups. Distribute markers, pencils, and a copy of Visual 12 to each group. Assign the consecutive numbers 101–200, 201–300, 301–400, and so on, to each group.

- Each group creates a number chart by filling in its assigned consecutive numbers on the grid.
- Group members use markers to circle all the numbers on their completed chart that are divisible by 4.
- Groups tape their completed charts on the chalkboard for all students to see. The class compares the circled numbers on the charts looking for patterns.

Customizing the Activity for Individual Needs

ESL Help students understand that *consecutive* means following one after another in order, without a break. Write the numbers 7, 2, 5, 9, 1, 8, 3, 10, 6, and 4 on the board. Ask pairs of students to rewrite the numbers in consecutive order.

Learning Styles Students can:

 look for patterns, such as how many numbers are circled in each row on the charts, or similarities in the positions the circled numbers occupy in each chart.

 work with counters to test divisibility by 4.

 use their own words to describe the divisibility test for 4.

Reinforcement Activity

Have students work in pairs. First one student writes a number (up to millions). The other student uses the divisibility tests to decide whether the number is divisible by 2, 3, 4, 5, 6, 9, 10, or not. Then students reverse roles.

Alternative Assessment

Students can determine if a number is divisible by 4 and explain their reasoning.

Example: Is 114 divisible by 4? How do you know? Is 136 divisible by 4? How do you know?

Answers: Accept all reasonable explanations. Students may say that 114 is not divisible by 4 because 14 is not divisible by 4; 136 is divisible by 4 because 36 is divisible by 4.

USING YOUR CALCULATOR
Test for Divisibility
Student Edition page 113

Lesson Objective
- Life Skill: Use a calculator to determine divisibility.

Game
What's It Divisible By?
Materials: index cards, calculators, colored markers
Procedure: Organize students into four groups. Provide index cards, markers, and calculators for each student. Tell students that they can use a calculator to test divisibility by any number. Write the following numbers on the chalkboard: 12, 18, 32, 48. Assign a different number to each group. Post the following general rule:

> Divisibility Test for "n"
> Divide a number by "n". If the quotient is a whole number, it is divisible by "n".

(Continued on p. 64)

(Continued from p. 63)

- Each student uses a calculator to make up divisibility cards for the assigned number. He or she multiplies the assigned number by any 2- or 3-digit number. The product is written on the front of an index card and the assigned number is written on the back.
- Group members then swap cards and test the products. Students divide each product by the assigned number. If the quotient is a whole number, the product is divisible by the assigned number. The card is correct.
- All the cards are collected and shuffled by a volunteer who then picks a card and calls out the product to the groups.
- Groups work together, using trial and error, to find the assigned number by which the product is divisible. The first group to call out correctly gets a point. Repeat until a team scores 5 points.

6.4 Factors and Greatest Common Factor
Student Edition pages 114–115

Prerequisite Skill
- Using basic division facts

Lesson Objectives
- Find the factors of a number.
- Find the greatest common factor.
- Life Skill: Make a seating plan.

Words to Know
factors, greatest common factor (GCF)

Cooperative Group Activity

Match Factors
Materials: index cards, markers

Procedure: Organize students into pairs. Give each pair 40 index cards and markers. Assign the number 12 to one member of the pair and the number 16 to the other.

- Each student numbers a set of index cards 1–20.
- Pairs work together finding all the factors of their assigned number. They place their factor cards face up on the table.
- They compare the factor cards matching those that are the same for both numbers. These are the common factors of 12 and 16. Students identify the greatest common factor.
- Repeat the activity using other numbers, such as 12 and 18, 15 and 18, 18 and 20.

Customizing the Activity for Individual Needs
ESL To help students understand what *factors* are, have pairs take turns writing multiplication facts and circling the factors in each. To help students understand *greatest common factor*, have them examine each word separately. Discuss other words or expressions that use *greatest* and *common*, such as "We have a *common* interest."

Learning Styles Students can:

 circle common factors with the same color.

 arrange cards showing common factors in order from least to greatest to find the greatest common factors, lining up the cards in the quotient with their proper place values.

 listen and follow the steps as another describes how to find the factors of a given number.

Enrichment Activity
Have students determine how many of each size dining room table would be needed for 120 people if only one type of table could be used at a time with no empty chairs: 8-person table; 10-person table; 12-person table; 15-person table.

Alternative Assessment
Students can find the factors of a number by using counters to make as many rectangles as possible.

Example: Find the factors of 6.

Answer: $F_6 = \{1, 2, 3, 6\}$ Students make four arrangements of counters as shown below.

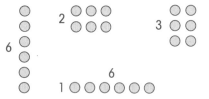

6.5 Multiples and Least Common Multiple
Student Edition pages 116–117

Prerequisite Skills
- Using basic division facts
- Using basic multiplication facts

Lesson Objectives
- Find the multiples of a number.
- Find the least common multiple.
- Life Skill: Plan appointments on a calendar.

Words to Know
multiples, least common multiple (LCM)

Cooperative Group Activity

Name That Multiple
Materials: paper strips, empty boxes, pencils

Procedure: Assign students to groups of two to four people. Give each group 12 strips of paper and an empty box.
- Each group of students writes the numbers 5 to 16, one number per strip. Students put the strips in the empty box.
- Group members take turns selecting two strips from the box, writing down the numbers. They then name one common multiple for the two numbers picked.
- Other students in each group take turns naming other common multiples for the two numbers up to and including their product. Each student who names a common multiple scores 1 point. The student who names the least common multiple scores 2 points.
- Students return the strips to the box and repeat the activity until one group member scores 10 points to win.

Customizing the Activity for Individual Needs
ESL Demonstrate the difference between *multiple* and *multiply*, using a multiplication table. Point out the row with the *multiples* of 4. To find the *multiples* of 4, you *multiply* each whole number, 0 through 9, by 4. A *multiple* is a number. To *multiply* is an activity.

Learning Styles Students can:

 circle common multiples.

 write multiples on separate index cards and arrange the cards in order from least to greatest.

 say the multiples of a number aloud as they generate a written list of multiples. They say their lists of multiples aloud when comparing lists to find the LCM.

Reteaching Activity

Students can use a number line or Visual 1 *Number Lines: Whole Numbers* to find the least common multiple of 3 and 4. Have them skip count to identify the first five multiples of 3. Then skip count to find the multiples of 4. They stop when the multiple of 4 is also a multiple of 3.

The least common multiple of 3 and 4 is 12. Repeat with different pairs of numbers.

Alternative Assessment

Students can identify a missing multiple.

Example: Identify the missing multiple in this list of numbers: 6, 12, ?, 24, 30   Answer: 18

6.6 Prime Numbers
Student Edition page 118

Prerequisite Skill
- Finding factors of a number

Lesson Objectives
- Identify prime and composite numbers.
- Write the prime factorization of a number.

Words to Know
prime number, composite number

Cooperative Group Activity

Making Factor Trees
Materials: number cubes, pencils, paper

Procedure: Organize students into small groups and give each group a pair of number cubes, paper, and pencils.
- Group members take turns tossing both number cubes to generate a 2-digit number.
- After each toss, students work in pairs making factor trees to find the prime factorization of the number.
- Pairs compare the steps in each of their prime factorizations. They see that no matter how they factor any one number, the prime factorization is the same.
- Repeat until all have a turn tossing.

Customizing the Activity for Individual Needs

ESL Help students differentiate between *prime* and *composite numbers*. Discuss the meanings of *prime* as it is related to *primary* (basic, original). Such as the *primary source*. Then brainstorm words that are similar to *composite*—all should involve the putting together or creation of something from other parts, such as a *composer* of music and a written *composition*.

Learning Styles Students can:

 use different colors for the branches of the prime factorization "trees."

 arrange tiles in as many rectangles as possible to factor a number. More than two arrangements means *composite*; only two arrangements mean *prime*.

 alternate explaining and listening to other students as they describe how to find the prime factorization of a number.

Enrichment Activity

Have students use a factor chart and a calculator to find the prime factorization of a number. Start with division by 2 as the first prime factor. If the quotient is a whole number, write it in the quotient column. Bring it into the next row of the number column. Try 2 again. If the quotient is not a whole number, mark a slash and start a new row. Try the next prime number with the same factor. Continue trying primes 3, 5, 7, … until the quotient is 1. Write the factorization, using the prime factors. Multiply to check.

Example: Find the prime factorization of 50.

Factor Chart				
Number		Prime Factor		Quotient
50	÷	2	=	25
~~25~~	~~÷~~	~~2~~	~~=~~	\
~~25~~	~~÷~~	~~3~~	~~=~~	\
25	÷	5	=	5
5	÷	5	=	1

Answer: $2 \times 5 \times 5 = 50$

Alternative Assessment

Students can identify prime numbers:

Example: Find any 3 prime numbers on the hundreds chart.

Answer: Any three primes: 2, 3, 5, 7, 11, 13, 17, 19, 23, 29, 31, 37, 41, 43, 47, 53, 59, 61, 67, 71, 73, 79, 83, 89, 97.

6.7 Exponents
Student Edition page 119

Prerequisite Skill
- Knowing basic multiplication facts

Lesson Objective
- Find the value of expressions containing exponents.

Words to Know
exponent

Cooperative Group Activity

Exponent Cards

Materials: index cards of two colors, markers

Procedure: Organize students into groups of three. Give each group ten index cards of one color and five index cards of the second color.

- Students write the numbers 0–10 in large print on the first set of cards, one number to a card. These cards will be used as factors.
- Students write the numbers 0–4 on the second set of cards in smaller print. These cards will be used as exponents.
- Students mix up the order of the cards in each set and place the cards face down on their desks in two piles.
- One student in each group picks a card from each pile and writes the exponential expression. Another student from the group reads the exponential expression out loud. The third student from the group evaluates the exponential expression.
- Students switch roles and repeat the activity.

Customizing the Activity for Individual Needs

ESL To explain the word *exponents*, point out that exponents are shorthand for writing repeated multiplication. Tell students that the *x* in ex*x*ponent reminds them to multiply.

Example:
$2^5 = 2 \times 2 \times 2 \times 2 \times 2$

66 • Chapter 6

Learning Styles Students can:

 use separate colors for writing factors and exponents in exponential expressions. They then repeat the color used for the factor as they evaluate the expressions.

 arrange number cards to form and evaluate exponential expressions.

 say exponential expressions in words. "Two to the third power is two times two times two, which equals eight."

Reinforcement Activity

Have students use number cards to model different exponential expressions. For example, for the expression 10^3, students use number cards to arrange the factors: $10 \times 10 \times 10$.

Alternative Assessment

Students can write the exponential expressions for a repeated multiplication problem.

Example: Write the exponential expression for $3 \times 3 \times 3 \times 3$.

Answer: 3^4

6.8 Squares and Square Roots
Student Edition page 120

Prerequisite Skill
- Using exponents

Lesson Objective
- Find squares and square roots.

Words to Know
square, square root

Cooperative Group Activity

Matching Roots and Squares

Materials: index cards, markers

Procedure: Divide the class into two teams: Roots and Squares. Give each team 15 index cards. Have the Roots team write the numbers 1–15 on separate cards. Have the Squares team write the squares of the numbers 1–15 on separate cards.

- A member of the Roots team holds up one of the cards. The Squares team finds the matching square and holds it up.
- One member of each team comes up to the chalkboard to write a root or square problem that relates the two numbers. For example, if the Roots team holds up a 3, the Squares team holds up a 9; the member of the Roots team would write $3 = \sqrt{9}$; the member of the Squares team would write $9 = 3^2$.
- Students put aside the used cards and repeat the activity starting with a member of the Squares team. Teams alternate starting until all the cards are used up.

Customizing the Activity for Individual Needs

ESL To help students understand the relationship between the mathematical term *square* and the geometric shape, point out that the sides of a square are equal. Use tiles to create a square shape. Have students count the number of tiles on each side and the total number of tiles. Point out that one side of the square represents the square root.

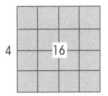

Learning Styles Students can:

 shade squares on grid paper to find the square of a number. They can count the number of shaded squares on one side to find the square root of the number.

 use number cards to create pairs of matching squares and square roots.

 say the squares and square roots aloud.

Reinforcement Activity

Have students use the multiplication table to find all the square numbers between 1 and 81. Have them make a table of these squares and their square roots on an index card for easy reference.

Alternative Assessment

Students can find the square root of a number by using tiles to form a square and counting the number of tiles on a side.

Example: Use 9 tiles to find $\sqrt{9}$.

Answer: 3; students arrange the tiles in a square with 3 tiles on each side.

ON-THE-JOB MATH
Electric Meter Reader
Student Edition page 121

Lesson Objectives
- Apply number sense to reading electrical meters.
- Life Skill: Calculate the cost of using electricity and interpret a bill.
- Communication Skill: Write a letter.
- Workplace Skill: Read a meter and estimate future electricity usage.

Activities
Look It Up
Materials: electric bills with meter readings
Procedure: Have students circle and label the following sections on a bill:
- the meter number
- the meter reading for that month
- the previous meter reading
- the kilowatt-hours used that month
- the cost per kilowatt-hour
- the cost for electricity for the month
- the next meter-reading date

Write a Business Letter
Materials: electric bill or phone book, paper and pen or computer word processor
Procedure: Have students write a letter to the local electric company. They can find the address on an electric bill or in the telephone book. Have students ask for information about becoming an electric meter reader, including the type of training needed.

Act It Out
Materials: none
Procedure: Have two students role-play being an electric meter reader and customer or manager. Give them the following scenarios:
- Ms. Morales is an electricity customer. She has electric heat. She also has many lamps and electrical appliances. Her electric bills are very high. She asks the meter reader what she can do to lower her bill.
- The following readings were taken on Mr. Smith's electric meter. March 1: 15,234 kilowatt-hours. April 1: 16,739 kilowatt-hours. May 1: 17,504 kilowatt-hours. In June, Mr. Smith was away and you couldn't read the meter. You must record an estimated amount of kilowatt-hours. Explain how you figured the estimate.

Practice
Have students complete On-The-Job Math 73 *Electric Meter Reader* from the Classroom Resource Binder.

Internet Connection
Have students do a search for and visit the Web sites of local or nearby utility companies. There they will find employment information as well as consumer information on saving electricity, reading meters, and understanding electric bills. Consumer information about meter reading can also be found on the Web sites of many state Department of Public Services. One such site is:
New York Department of Public Services
http://www.dps.state.ny.us/meter.html

6.9 Problem Solving: Extra Information
Student Edition pages 122–123

Prerequisite Skill
- Solving word problems with whole numbers

Lesson Objectives
- Solve word problems with extra information.
- Solve different problems using the same situation.
- Life Skill: Calculate earnings and pay.

Cooperative Group Activity

Problems with Something Extra
Materials: pencils, paper

Procedure: Organize students into groups of four to write problems. Assign one student in each group an addition problem, a subtraction problem, a multiplication problem, and a division problem. Encourage students to write problems based on everyday experiences.
- On separate sheets of paper, groups work together to write word problems containing extra information. On the back of each paper, they identify the extra information and the solutions.
- Students exchange problems with other groups, circle the extra information and solve.
- Students check each other's answers.

Customizing the Activity for Individual Needs
ESL Help students identify the words and numbers needed to solve the problems. Have them

draw a box around the key words in the problem related to each question. They can circle the numbers and units related to the key words they boxed. Finally, they cross out any extra information not related to the key words.

Learning Styles Students can:

 cross out extra information that is not needed to solve a problem and underline the information that is needed to solve a problem.

 write each part of the problem on a separate index card. Then remove the card with the information that is not needed.

 read the word problems out loud stressing key words and whispering extra information.

Enrichment Activity

Give students the situation below and ask them to write one or more problems based on all or some of the information.

There are 120 calories in 8 ounces of low-fat milk. There are 100 calories in 4 ounces of orange juice. Tim has a glass that holds 6 ounces.

Alternative Assessment

Students can identify and eliminate extra information from a word problem.

Example: Which sentence or sentences will not help you solve this problem?

Johanna is 16 years old. She earns $5 an hour baby-sitting. She baby-sat for 4 hours on Saturday night. How much did she earn?

Answer: Johanna is 16 years old.

Closing the Chapter
Student Edition pages 124–125

Chapter Vocabulary

Review with the students the Words to Know on page 107 of the Student Edition. Then have students quiz each other in pairs.

Have students copy and complete the Vocabulary Review questions on page 124.

For more vocabulary practice, have them complete Vocabulary 65 *More About Numbers* from the Classroom Resource Binder.

Test Tips

In pairs, students can take turns showing how to solve a problem by using one of the test tips.

Learning Objectives

Have the students review CM6 *Goals and Self-Check*. They can check off the goal they have reached. Note that each section of the quiz corresponds to a Learning Objective.

Group Activity

Summary: Students find different ways to arrange 135 members of a band in equal-sized rows.

Materials: pencils, paper

Procedure: Students can use counters to model the problem.

Assessment: Use the Group Activity Scoring Rubric on page xii of this guide. Fill in the rubric with the additional information below. For this project, students should have:

- listed eight possible arrangements.
- used all divisibility tests and answered the question.

RELATED MATERIALS See the unit overview page for other Globe Fearon books that can be used to enrich and extend the material in this unit.

Assessing the Chapter

Traditional Assessment

Chapter Quiz
The Chapter Quiz on pages 124–125 of the Student Edition can be used as either an open- or closed-book test, or as homework. Use the quiz to identify concepts in the chapter that students need to review and practice.

Chapter Tests
Use Chapter Test A Exercise 75 and Chapter Test B Exercise 76 *More About Numbers,* from the Classroom Resource Binder to further assess mastery of chapter concepts.

Additional Resources
Use the Resource Planner on page 60 to assign additional exercises from the program.

Alternative Assessment

Interview
Present the number 12 and ask:
- Is 12 divisible by 2? by 3? by 4? by 5? by 6? by 9? by 10?
- What are the factors of 12?
- What are the first five multiples of 12?
- Is 12 prime or composite?
- What is the prime factorization of 12?
- What is 12^2?

Presentation
Write the numbers 75 and 100 on the chalkboard. Have students demonstrate how they would:
- test for divisibility by 3 and by 5.
- find their greatest common factor.
- find their least common multiple.
- tell which number is a square.

Unit Assessment

This is the last chapter in Unit 1, *Whole Numbers*. To assess cumulative knowledge and provide standardized-test practice, administer the practice test on page 126 of the Student Edition and Unit 1 Cumulative Test, page T7 in the Classroom Resource Binder. These tests are in multiple-choice format. A scantron sheet is provided on page T2 of the Classroom Resource Binder.

Unit Overview

Unit 2 ▶ Fractions

CHAPTER 7 Fractions and Mixed Numbers	**PORTFOLIO PROJECT** Fraction and Mixed Number Journal	**MATH IN YOUR LIFE** Cooking	**USING YOUR CALCULATOR** Comparing Fractions	**PROBLEM SOLVING** Patterns
CHAPTER 8 Multiplying and Dividing Fractions	**PORTFOLIO PROJECT** Recipe	**ON-THE-JOB MATH** Car Rental Agent	**USING YOUR CALCULATOR** Multiplying Fractions and Whole Numbers	**PROBLEM SOLVING** Solve a Simpler Problem Does the Answer Make Sense?
CHAPTER 9 Adding and Subtracting Fractions	**PORTFOLIO PROJECT** Activities	**MATH IN YOUR LIFE** Pay Day	**USING YOUR CALCULATOR** Finding Common Denominators	**PROBLEM SOLVING** Multi-Part Problems

Help your students succeed in math mastery with this consistent, success-based program.

Enrich your math test-taking curriculum with this complete writing text that shows your student the importance of communicating clearly.

Teach essential everyday math skills through authentic, real-life activities.

RELATED MATERIALS

These are some of the Globe Fearon books that can be used to enrich and extend the material in this unit.

Practice & Remediation ▶

Simple Fractions
Help your students succeed in math mastery with this consistent, success-based program.

Test Preparation ▶

Writing In Mathematics
Enrich your math test-taking curriculum with this complete writing text that shows your student the importance of communicating clearly.

Real-World Math ▶

Using Dollars and Sense
Teach essential everyday math skills through authentic, real-life activities.

Planning the Chapter

Chapter 7 • Fractions and Mixed Numbers

Chapter at a Glance

SE page	Lesson	
128		Chapter Opener and Project
130	7.1	What Is a Fraction?
131	7.2	Recognizing Equivalent Fractions
132	7.3	Reducing Fractions to Lowest Terms
134	7.4	Changing Fractions to Higher Terms
136	7.5	Finding Common Denominators
138	7.6	Comparing Fractions
140	7.7	Ordering Fractions
141		Using Your Calculator: Comparing Fractions
142	7.8	Changing Fractions to Mixed Numbers
144	7.9	Changing Mixed Numbers to Fractions
145		Math in Your Life: Cooking
146	7.10	Ordering Numbers You Know
148	7.11	Problem Solving: Patterns
150		Chapter Review and Group Activity

Learning Objectives
- Identify fractions and their parts.
- Recognize equivalent fractions.
- Write fractions in lowest terms and in higher terms.
- Find common denominators.
- Compare and order fractions and other numbers.
- Change mixed numbers and fractions.
- Solve problems using patterns.
- Apply fractions to cooking.

Life Skills
- Keep a journal.
- Compare fractions in everyday situations, such as eating pizza, shopping, and reading signs.
- Use a calculator to compare fractions.
- Adapt recipes.
- Use equivalent fractions to measure recipe ingredients.
- Compare stock prices.

Communication Skills
- Read fractions and mixed numbers on signs and in newspapers, magazines, supermarkets, and recipes.
- Explain how to use equivalent measures.
- Write a letter.

PREREQUISITE SKILLS
To assess mastery of prerequisite skills, use a selection of exercises from any of the program resources referenced below. The same resources can be used to provide remediation if necessary.

Program Resources for Review

Skills	Student Edition	Workbook	Classroom Resource Binder
Counting whole numbers	Lesson 1.1	Exercise 1	
Ordering whole numbers	Lesson 1.7	Exercise 7	Practice 6 Mixed Practice 8
Adding whole numbers	Lesson 2.2	Exercise 11	Review 14, 15 Practice 17 Mixed Practice 18

Program Resources for Review

Skills	Student Edition	Workbook	Classroom Resource Binder
Multiplying whole numbers	Lesson 4.2	Exercise 27	
Dividing whole numbers	Lesson 5.2 Lesson 5.3	Exercise 36 Exercise 37	Review 50
Finding the greatest common factor of two numbers	Lesson 6.4	Exercise 48	Review 66 Practice 70
Finding the least common multiple of two numbers	Lesson 6.5	Exercise 49	Review 67 Practice 71

Diagnostic and Placement Guide

Chapter 7 Fractions and Mixed Numbers

The Percent Accuracy scores are based on the number of problems in a lesson that have been answered correctly.

Resource Planner

After diagnosing your students' needs, use the correlating Program Resources for reinforcement, reteaching, or enrichment. Additional activities for customizing the lessons can be found in this guide.

KEY
Reteaching = ⤴   Reinforcement = ⬇   Enrichment = ⤴

Lessons		Percent Accuracy				Program Resources			
		50%	65%	80%	100%	Student Edition Extra Practice	Workbook Exercises	Teacher's Planning Guide	Classroom Resource Binder
7.1	What Is a Fraction?	1	2	2	3		54	⤴ p. 75	
7.2	Recognizing Equivalent Fractions	4	5	6	8		55	⬇ p. 75	
7.3	Reducing Fractions to Lowest Terms	20	26	32	40	p. 422	56	⤴ p. 76	⬇ Review 78 *Writing Fractions in Lowest Terms and Higher Terms*
7.4	Changing Fractions to Higher Terms	12	16	19	24	p. 422	57	⤴ p. 77	⬇ Review 78 *Writing Fractions in Lowest Terms and Higher Terms*
7.5	Finding Common Denominators	15	20	24	30	p. 422	58	⬇ p. 78	⬇ Review 79 *Finding Common Denominators*
7.6	Comparing Fractions	11	14	17	21		59	⤴ p. 79	⬇ Review 80 *Comparing Fractions*
7.7	Ordering Fractions	9	12	14	18		60	⤴ p. 79	
7.8	Changing Fractions to Mixed Numbers	10	13	16	20		61	⤴ p. 81	⬇ Practice 82 *Changing Fractions to Mixed Numbers, Changing Mixed Numbers to Fractions*
7.9	Changing Mixed Numbers to Fractions	6	8	10	12	p. 422	62	⤴ p. 81	⬇ Practice 82 *Changing Fractions to Mixed Numbers, Changing Mixed Numbers to Fractions*

Lessons	Percent Accuracy				Program Resources			
	50%	65%	80%	100%	Student Edition Extra Practice	Workbook Exercises	Teacher's Planning Guide	Classroom Resource Binder
Math in Your Life: Cooking	Can be used for portfolio assessment.						p. 82	Math in Your Life 83 Cooking
7.10 Ordering Numbers You Know	7	9	11	14		63	p. 83	Challenge 84 Working with Fractions
7.11 Problem Solving: Patterns	1	2	2	3		64	p. 83	Review 81 Problem Solving: Patterns
Chapter 7 Review								
Vocabulary Review	6.5	8.5	10.5	13			p. 84	Vocabulary 77
	(writing is worth 4 points)							
Chapter Quiz	13	16	20	25			p. 85	Chapter Test A, B 85, 86

Customizing the Chapter

Opening the Chapter
Student Edition pages 128–129

Photo Activity

Procedure: Discuss the stock market. Point out that the price of a stock is in dollars and part of a dollar. A stock that is $18\frac{1}{4}$ is selling at $18\frac{1}{4}$ or $18.25. The smallest fractional part on most major stock exchanges is $\frac{1}{8}$ of a dollar or $12\frac{1}{2}$ cents.

Provide students with examples of financial pages from a newspaper and have them practice reading mixed numbers aloud. Have each student select two or three stocks they can track over the course of the next few weeks. Have them find out if the stock value went up or down.

Words to Know

Review with students the Words to Know on page 129 of the Student Edition. Help students remember words related to the parts of a fraction by having students write a fraction and label the parts.

The following words and definitions are covered in this chapter.

fraction a form of number that shows part of a whole

numerator the top number in a fraction

denominator the bottom number in a fraction

equivalent fractions fractions with different numbers but equal values

lowest terms when only 1 divides evenly into both the numerator and denominator of a fraction

like fractions fractions with the same denominator

unlike fractions fractions with different denominators

mixed number a number made up of a whole number and a fraction

Fraction and Mixed Number Journal Project

Summary: Students keep a daily journal, listing and describing fractions and mixed numbers they see on containers, signs, newspaper, magazines, and other places.

Materials: newspapers, magazines, containers, almost any print material; paper, pencils

Procedure: Provide students with examples of research material, such as newspapers and magazines. Students can complete this project in class with all materials provided. Or, students can keep a journal throughout this chapter and cite examples found during their regular daily routine. Students may choose a journal entry they like for their portfolios.

Assessment: Use the Individual Activity Rubric on page *xi*. Fill in the rubric with the additional information below. For this project, students should have:

- found at least three fractions and three mixed numbers.
- identified where and how each fraction or mixed number is used.

Learning Objectives

Review with the students the Learning Objectives on page 129 of the student edition before starting the chapter. Students can use the list of objectives as a learning guide. Suggest that they write the objectives in a journal or use *Goals and Self-Check* CM6.

After each lesson, have students write an example of the skill they learned under the appropriate objective. Suggest that students use these notes as a learning guide to help them study for the chapter test.

7.1 What Is a Fraction?
Student Edition page 130

Prerequisite Skill
- Counting whole numbers

Lesson Objective
- Identify fractions and their parts.

Words to Know
fraction, numerator, denominator

Cooperative Group Activity

Fraction Toss
Materials: paper bags, two-color counters or coins, paper, pencil

Procedure: Prepare bags of counters with different numbers of counters in each bag. Organize students in pairs. Give each pair a bag of counters.

- One student tosses the counters from the bag onto the desk. The student names the fraction of counters that lands with one particular color (or heads or tails) showing, and writes the fraction. The other student tells the fraction of counters that lands with the other color showing, then writes the fraction.
- Students take turns tossing the counters on the table.
- To vary the game, one student writes a fraction. The other uses the counters to model the fraction.

Customizing the Activity for Individual Needs

ESL Student pairs help each other pronounce fraction names, paying special attention to the /th/ sound at the end of most names, such as one-four*th* and three-sixtee*nths*.

Learning Styles Students can:

 illustrate fractions on the board.

 use two-color counters or coins to model fractions.

 say the fractions aloud after the two-color counters have been tossed.

Reteaching Activity

Provide each student with fraction bars showing halves, thirds, fourths, fifths, sixths, eighths, tenths, and twelfths or use Visual 8 *Fraction Bars*. Ask students to use shading to show each of the following fractions:

$\frac{1}{2}, \frac{2}{3}, \frac{3}{4}, \frac{2}{5}, \frac{1}{6}, \frac{5}{8}, \frac{3}{10}, \frac{7}{12}$

Alternative Assessment

Students can recognize a fraction as part of a whole or part of a set.

Example: Use shading to show the meaning of each fraction below.

  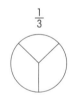

Answer: 7 out of 12 parts are shaded; 3 out of 4 circles are shaded; $\frac{1}{3}$ of the circle is shaded.

7.2 Recognizing Equivalent Fractions
Student Edition page 131

Prerequisite Skill
- Identifying fractions and their parts

Lesson Objective
- Recognize equivalent fractions.

Words to Know
equivalent fractions

Chapter 7 • 75

Cooperative Group Activity

Folding Fractions

Materials: paper, crayons or colored pencils

Procedure: Organize students into groups. Give each student two sheets of paper and a crayon or colored pencil. Have students begin by folding the paper in half and shading one of the halves, keeping the paper folded. Ask students how many parts of the paper are shaded.

- Students name and write the fraction telling what part of the paper is shaded.
- Students fold the paper in half again, then unfold it all the way. They count how many parts of the paper are shaded. Then they name and write the fraction telling what part is shaded. A volunteer explains how he or she knows the two fractions are equivalent.
- Students refold the paper twice, then fold it in half one more time. They unfold the paper all the way and tell how many parts are shaded. They write another equivalent fraction.
- *Variation*: Students fold another sheet of paper into three equal parts and color in one third. They repeat the activity by folding the paper in half to find equivalent fractions for $\frac{1}{3}$. They see how many times they can fold the paper to get equivalent fractions.

Customizing the Activity for Individual Needs

ESL To reinforce understanding of *equivalent fractions*, make equivalent-fraction flash cards that show fractions on one side. First, have students match pairs of equivalent fractions. Then hold up one card, and have students write and say the equivalent fractions for the fraction shown.

Learning Styles Students can:

 write strings of equivalent fractions, connecting them with equal signs.

 cut out shaded shapes and place them one on top of another to test whether fractions are equivalent.

 explain why the fractions are equivalent.

Enrichment Activity

Provide construction paper, scissors, and markers. Have students cut out and fold sheets of paper to make fraction models with other shapes, such as circles, triangles, and so on. Encourage them to refold a model to find and label as many fractions as they can that are equivalent.

Examples: Start with $\frac{3}{4}$, folding in half gives $\frac{6}{8}$; folding in thirds gives $\frac{9}{12}$; folding in fourths again gives $\frac{12}{16}$.

Have students use the equivalent fraction displays for a bulletin board.

Alternative Assessment

Students can recognize equivalent fractions using fraction bars or Visual 8 to show halves, thirds, fourths, fifths, sixths, eighths, tenths, and twelfths.

Example: Find a fraction that is equivalent to $\frac{1}{3}$.

Answer: $\frac{2}{6}$ or $\frac{4}{12}$

7.3 Reducing Fractions to Lowest Terms
Student Edition pages 132–133

Prerequisite Skills

- Finding greatest common factors
- Dividing by whole numbers

Lesson Objectives

- Reduce fractions to lowest terms.
- Life Skill: Compare serving sizes of food.

Words to Know

lowest terms

Cooperative Group Activity

Rolling Fractions

Materials: blank number cubes, paper, pencils, transparent markers

Procedure: Organize students into groups of two or three. Give one pair of number cubes, paper, pencils, and a transparent marker to each group. Assign different multiples to each group. For example, one group might use 3, 6, 9, 12, 15, and 18; another group might use 4, 8, 12, 16, 20, and 24.

- Students write their assigned multiples on each number cube.
- Group members take turns tossing the number cubes. They write a fraction using the smaller number as the numerator. Students work together to reduce the fraction to lowest terms. One member of the group records the results.

- After several tosses, groups exchange papers to check each other's work. They then exchange number cubes and repeat the activity.

Customizing the Activity for Individual Needs

ESL To help students understand how the words *lowest terms* can be used in a sentence, write or say the following:

"The fraction $\frac{1}{2}$ is in *lowest terms*."

"I can reduce $\frac{4}{8}$ to *lowest terms*."

"The fraction $\frac{4}{8}$ in *lowest terms* is $\frac{1}{2}$."

Learning Styles Students can:

 draw models of the given fraction and its reduced form to make sure that they are equivalent.

 use fraction bars or Visual 8 to help reduce fractions to lowest terms.

 describe how to reduce each fraction to lowest terms.

Reteaching Activity

Provide students with three different-colored markers and circles divided into 2, 3, 4, 5, 6, 8, 10, and 12 equal sections or Visual 7 *Fraction Circles*. For each situation below, have students shade one circle to show what fraction each person ate. Students then shade a second circle to show the fraction in lowest terms. Encourage students to write a statement that tells why the two fractions are equivalent. Use a different color marker for each situation.

(a) Carla ate 5 sections of an orange that had 10 sections.

(b) Leon ate 4 slices of pizza from a pie that was cut into 6 equal pieces.

(c) Joe ate 3 pieces of pie from an apple pie that was cut into 12 equal pieces.

Alternative Assessment

Students can reduce a fraction to lowest terms by drawing a diagram.

Example: Reduce $\frac{4}{6}$ to lowest terms. Draw a diagram.

Answer: $\frac{2}{3}$; students should draw a diagram similar to the one shown below.

7.4 Changing Fractions to Higher Terms
Student Edition pages 134–135

Prerequisite Skills
- Using basic multiplication and division facts
- Multiplying and dividing by whole numbers

Lesson Objectives
- Change fractions to higher terms.
- Life Skill: Make varieties of equivalent food servings.

Cooperative Group Activity

Make It Bigger

Materials: fraction cards, number cubes, paper, pencils

Procedure: Have students work in pairs.

- One student picks a fraction card. The other student rolls a number cube.
- Pairs work together to change the fraction picked by multiplying numerator and denominator by the number rolled on the cube.
- They record the original fraction and the higher-term fraction together on an index card.
- Students reverse roles and repeat the activity.
- *Variation:* Students can use the index cards of equivalent fractions to make higher-term flash cards. Have them write the numerator of the higher-term fraction on the back of the index card. Then replace this numerator with a question mark on the front of the card. Students work in pairs to solve each puzzle. They check the back of the card for the correct answer.

Customizing the Activity for Individual Needs

ESL To help explain the phrase *higher terms*, relate to finding *lowest terms*. Explain and demonstrate finding both. Point out that with *lowest* terms you're dividing and writing *smaller* numerators and denominators. In *higher* terms, you're multiplying and the numbers are *larger*.

Learning Styles Students can:

 shade the *Fraction Bars* on Visual 8 to demonstrate how to find higher-term fractions.

 manipulate fraction bars to model a given fraction and its higher-term fraction.

 say aloud the number by which to multiply the numerator when the denominator is given.

Enrichment Activity

Have students make up word puzzles like the following for their classmates to solve.

I'm thinking of a fraction that is equivalent to $\frac{2}{3}$. The denominator is between 9 and 15. What is the fraction?

Answer: $\frac{8}{12}$

Alternative Assessment

Students can use a model to change a fraction to higher terms for a specific denominator.

Example: Show $\frac{3}{4}$ in higher terms. Use the denominator 12.

Answer: Students use fraction bars to show that $\frac{3}{4} = \frac{9}{12}$.

7.5 Finding Common Denominators
Student Edition pages 136–137

Prerequisite Skills
- Multiplying and dividing by whole numbers
- Finding the least common multiple

Lesson Objectives
- Find common denominators.
- Change unlike fractions to like fractions.
- Life Skill: Compare weights at the supermarket.

Words to Know
like fractions, unlike fractions

Cooperative Group Activity

Like and Unlike Fractions

Materials: Visual 9 *Fraction Cards*, pencils, paper

Procedure: Organize students into groups of three or four. Mix the fractions cards and spread them out randomly, face down on your desk.

- One student from each group chooses two fraction cards. The student writes down the fractions for the group to use.

- Group members work together to find the common denominator. They change the fractions to like fractions. They write the like fractions they found.

- After calculations are completed, another student from each group returns the fraction cards and select two others. Students repeat the activity until all group members have had a chance to choose fraction cards.

- Students from each group write the original fraction pairs and the like fractions their group found on the chalkboard. Other groups verify the results.

Customizing the Activity for Individual Needs

ESL Help students differentiate between *like fractions* and *unlike fractions*. Ask them to name some ways they are *like* their classmates and some ways they are *unlike* their classmates.

Learning Styles Students can:

 make separate lists of like and unlike fractions from all the fractions written on the chalkboard.

 organize fraction cards by common denominator.

 explain how to find common denominators and say the like fractions aloud.

Reinforcement Activity

Provide students with Visual 9 *Fraction Cards* and Visual 16 *Blank Table*. Have them play the *Like Fraction Match Game*. To set up the table, have them write the word *fraction* in the first row of the first column. Then write their names in the first row of the following columns. Students then write the fractions $\frac{1}{1}, \frac{1}{2}, \frac{1}{3}, \frac{1}{4}, \frac{1}{5}, \frac{1}{6}$ and $\frac{1}{8}$ in the first column. Have the students cut out the fraction cards; shuffle the cards; and place them face down. In turn, each player picks a fraction card and places it under their name in a row that is a like fraction. Students put aside cards they cannot use. The first player to fill in his or her column wins.

Fraction	Amy	Ling	Kai
$\frac{1}{1}$	$\frac{3}{1}$	$\frac{2}{1}$	
$\frac{1}{2}$	$\frac{5}{2}$	$\frac{6}{2}$	$\frac{3}{2}$
$\frac{1}{3}$		$\frac{4}{3}$	$\frac{5}{3}$
$\frac{1}{4}$	$\frac{3}{4}$	$\frac{2}{4}$	

Alternative Assessment

Students can explain how to change two unlike fractions to like fractions.

Example: Explain how you would change $\frac{1}{4}$ and $\frac{3}{8}$ to like fractions. What denominator would you choose? Why? What are the like fractions?

Answers: Use the denominator 8 because it is the LCM of 4 and 8. Like fractions are $\frac{2}{8}$ and $\frac{3}{8}$.

7.6 Comparing Fractions
Student Edition pages 138–139

Prerequisite Skills
- Finding common denominators
- Comparing whole numbers

Lesson Objectives
- Compare fractions.
- Life Skill: Compare measurements.

Cooperative Group Activity

Matching Denominators

Materials: Visual 9 *Fraction Cards*, paper, pencils

Procedure: Have students work in pairs. Mix the fractions cards and spread them out randomly, face down on your desk.
- Each student chooses a fraction card and writes down the fraction on a sheet of paper.
- Partners multiply the numerator and denominator of their respective fractions by 2, then by 3, by 4, and so on. Students list each multiple in order on their papers next to their original fractions.
- Partners compare their multiples each time until they see a common denominator.
- Students box the fractions with the common denominator and compare the numerators.
- Each student writes the inequality symbol that shows how their fractions compare.

Customizing the Activity for Individual Needs

ESL Relate the words *greater than* and *less than* to the symbols > and <. Show some examples using whole numbers such as 8 > 1 and 52 < 93. Give students other whole numbers to compare. Have them write the symbols to compare the numbers as they say the comparisons aloud.

Learning Styles Students can:

 write the words and the symbol when comparing fractions.

 line up shaded fraction bars under each other to compare fractions.

 say the number by which the numerator and denominator should be multiplied to make like fractions. They read their comparisons out loud.

Reteaching Activity

Provide fraction bars in 2, 4, 8, and 16 parts or Visual 8 *Fraction Bars*.

Ask students to show $\frac{3}{4}$ and $\frac{7}{8}$.

Have students line up the bars one under the other to see which is longer. Have them use symbols to describe the relationship between the two fractions, $\frac{3}{4} < \frac{7}{8}$. Repeat the activity using other fractions with denominators of 2, 4, 8, or 16.

Alternative Assessment

Students can compare fractions by naming a fraction greater than and a fraction less than a given fraction.

Example: What fraction is less than $\frac{1}{2}$? What fraction is greater than $\frac{1}{2}$?

Sample answers: $\frac{1}{3}$, $\frac{1}{4}$, or $\frac{2}{5}$ is less than $\frac{1}{2}$; $\frac{2}{3}$, $\frac{3}{4}$, or $\frac{4}{5}$ is greater than $\frac{1}{2}$.

7.7 Ordering Fractions
Student Edition page 140

Prerequisite Skills
- Finding common denominators
- Ordering whole numbers
- Comparing fractions

Lesson Objective
- Order fractions from least to greatest.

Cooperative Group Activity

That's an Order!

Materials: fraction cards or Visual 9 *Fraction Cards*, index cards, colored pencils

Procedure: Organize students into groups of three. Mix the fractions cards and spread them out randomly, face down, on your desk. Give each student in a group a different-colored pencil.

Each student chooses a fraction card and then lists 10 multiples of the denominator of the fraction. Group members compare their lists and identify the least common multiple. This will become the least common denominator of the like fractions.

- Each student then changes his or her fraction to an equivalent fraction with this new denominator. Group members check to see that they now have three like fractions.
- They then compare the numerators of the like fractions and order the fractions from least to greatest.
- Have the recorder of the group write the like fractions from least to greatest. Then write the original fractions from least to greatest.
- Students repeat the activity.

Customizing the Activity for Individual Needs

ESL Point out that the ending *-er,* as in grea*ter,* is used when comparing two things. The ending *-est,* as in grea*test,* is used when comparing three or more things. Have students say and write comparisons using these endings. For example: "I am older than my sister" and "I am the oldest of the three children in my family."

Learning Styles Students can:

 highlight the numerators of the like fractions to help order the fractions.

 arrange fraction bars by size to order fractions.

 say the multiples out loud until a common denominator is found.

Reteaching Activity

Have students use fraction bars to order the fractions $\frac{1}{4}$, $\frac{3}{16}$, and $\frac{5}{8}$ from least to greatest. Have them place the fraction bars under each other. Then have them rearrange the bars in size order with the smallest shaded part on top. Ask students to write the fractions in order.

Alternative Assessment

Students can explain in a guided interview the steps they would follow to order three fractions.

Example: Explain how you would order the fractions $\frac{2}{3}$, $\frac{5}{6}$, and $\frac{1}{2}$ from least to greatest.

How would you change the fractions? What denominator would you use? Why? Rewrite the fractions. Then order them.

Answer: Make them like fractions; 6 because it is the least common denominator; $\frac{4}{6}$, $\frac{5}{6}$, $\frac{3}{6}$; $\frac{3}{6}$, $\frac{4}{6}$, $\frac{5}{6}$

USING YOUR CALCULATOR
Comparing Fractions
Student Edition page 141

Lesson Objective
- Life Skill: Use a calculator to compare fractions.

Activities

Comparing Fraction Magic Numbers
Materials: calculators, paper, pencils
Procedure: Students can compare fraction "magic numbers" by turning fractions into whole numbers. They can divide numerator by denominator, multiply by 100, and compare the resulting magic numbers. They should ignore any numbers after the decimal point.

Show students how to compare $\frac{2}{5}$ and $\frac{1}{4}$ by following these steps:

Step 1 Divide numerator by denominator and then multiply by 100.

$$2 \div 5 \times 1 0 0 = \quad 40$$
$$1 \div 4 \times 1 0 0 = \quad 25$$

The magic number for $\frac{2}{5}$ is 40.
The magic number for $\frac{1}{4}$ is 25.

Step 2 Compare the magic numbers.
$$40 > 25$$

Step 3 Replace the magic numbers in the comparison with the fractions.
$$\frac{2}{5} > \frac{1}{4}$$

Have students compare the following fractions.

$\frac{5}{8}$ and $\frac{9}{10}$ $\frac{15}{18}$ and $\frac{5}{8}$ $\frac{4}{7}$ and $\frac{5}{9}$

Greater Than, Less Than, or Equal?
Materials: fraction cards; symbols cards for <, >, and =; calculator
Procedure: Have students play a game in pairs. One player randomly picks two cards from a pile of face-down fraction cards. The other

player arranges the fractions using the symbol cards to predict a comparison. Players use a calculator to make magic numbers and compare the fractions. A correct prediction scores 1 point. After 10 rounds, the player with the most points wins.

7.8 Changing Fractions to Mixed Numbers
Student Edition pages 142–143

Prerequisite Skills
- Dividing whole numbers
- Writing fractions in lowest terms

Lesson Objective
- Change fractions to mixed or whole numbers.

Words to Know
mixed number

Cooperative Group Activity

Pizza Fraction Challenge
Materials: paper, pencils

Procedure: Organize students into small groups. Assign one group the number 24 and the other groups any number from 25 to 50. Explain that these numbers represent the number of pizza slices they have. Draw a whole pizza cut into 6 slices on the chalkboard. Tell students that each of their pizzas will have 6 slices.

- Each group determines how many pies can be formed from the assigned number of slices. They also determine the number of slices that will be left over. Have them write the number of slices leftover as a fractional part of the whole pizza pie.
- A volunteer from each group shows the work that was done at the chalkboard.
- Students repeat the activity for a pizza that will have 8 slices.

Customizing the Activity for Individual Needs
ESL Help students differentiate between *whole numbers* and *mixed numbers*. Write the following whole numbers on the board: 1, 9, and 20. Underneath write: $1\frac{3}{4}$, $9\frac{3}{7}$, $20\frac{2}{3}$. Ask students what makes them different.

Learning Styles Students can:

 highlight the whole number part in one color and the fraction part in another color.

 cut up fraction bars to model improper fractions. Then reassemble the parts as mixed numbers.

 explain the work done by the group as another student writes it on the chalkboard.

Reteaching Activity
Provide students with several copies of circles divided into 2, 3, 4, 5, 6, 8, 10, and 12 equal sections or Visual 7 *Fraction Circles*. Have students cut apart the wedge shapes, keeping same-sized wedges together. Students then work in pairs. Each pair takes a handful of one size wedges and writes an improper fraction for the number of wedges. Then pairs put the wedges together to make as many whole circles as possible. Have students write down the whole or mixed number they find.

Alternative Assessment
Students can change a fraction to a mixed number using fraction pieces.

Example: Each fraction piece is $\frac{1}{4}$. There are 6 pieces. What is the mixed number?

Answer: The mixed number is $1\frac{2}{4}$ or $1\frac{1}{2}$.

7.9 Changing Mixed Numbers to Fractions
Student Edition page 144

Prerequisite Skills
- Adding whole numbers
- Multiplying whole numbers

Lesson Objective
- Change mixed numbers to fractions.

Cooperative Group Activity

Mixed Number Scramble
Materials: number cubes, pencils, paper

Procedure: Organize students into groups of three. Give each student a number cube.

- Group members all toss their number cubes at the same time. The group uses the resulting digits to make mixed numbers. They use one digit for the whole number, one digit for the numerator of the fraction, and one digit for the denominator of the fraction.

- Students work together to change each mixed number to a fraction.

Customizing the Activity for Individual Needs

ESL To help students understand the words in this chapter, have them make self-help cards with the word and an example. For example,

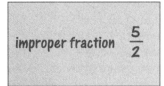

Learning Styles Students can:

 highlight each part of the mixed number in a different color.

 use fraction circles to represent mixed numbers. Then cut out shaded pieces and count them to find the improper fraction.

 say the mixed numbers and describe the procedure for changing to an improper fraction.

Enrichment Activity

Have students toss three number cubes and use the digits resulting from the toss to make the largest mixed number possible, the smallest number possible, and one number between the other two. Instruct students to make sure the fraction part of the mixed number is a proper fraction.

Alternative Assessment

Students can change a mixed number to a fraction using fraction pieces.

Example: Here are 2 whole circles and $\frac{3}{4}$ of a circle. What is the mixed number? Change the mixed number to a fraction.

Answer: The mixed number is $2\frac{3}{4}$. Students divide each whole circle into fourths, then count the number of fourths. The fraction is $\frac{11}{4}$.

MATH IN YOUR LIFE
Cooking Student Edition page 145

Lesson Objectives
- Apply fractions to cooking.
- Life Skill: Measure ingredients for a recipe.
- Life Skill: Compare measuring utensils.
- Communication Skill: Write a letter to request a recipe.

Activities

Tools of the Trade

Materials: cookbooks, paper, pencils

Procedure: Have students find a recipe in a cookbook and list the measurements of each ingredient from least to greatest (cups in one list, tablespoons in another list, and so on).

Have students take a trip to a store that sells kitchenware to see the kinds of measuring utensils that are sold. Ask students to compare the utensils used to measure wet ingredients with those used to measure dry ingredients.

Write a Business Letter

Materials: paper and pen or computer, food section of newspaper or food magazine

Procedure: Have students write a letter to the editor in charge of the food section of a newspaper or magazine, or to the host of a television cooking show. Tell them to:
- request a recipe for a type of food they like.
- ask how math is used in cooking.
- seek advice on kitchen safety measures.

Act It Out

Materials: measuring cups, spinach dip ingredients

Procedure: Have students bring in a set of measuring cups and the ingredients needed to make spinach dip. Take away one of the measuring cups ($\frac{1}{8}$, $\frac{1}{4}$, or $\frac{1}{2}$) from each group. Have students make the recipe using the remaining measuring cups.

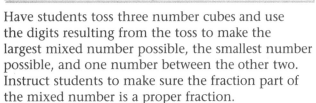

Practice

Have students complete Math in Your Life 83 *Cooking* from the Classroom Resource Binder.

Internet Connection

The following Web sites can be used to locate recipes, look up cooking terms, and find out about measuring utensils used in cooking:

Food Network
http://www.foodtv.com
digitalchef
http://www.digitalchef.com

7.10 Ordering Numbers You Know *Student Edition pages 146–147*

Prerequisite Skills
- Ordering whole numbers
- Ordering fractions

Lesson Objectives
- Order fractions, whole numbers, and mixed numbers.
- Life Skill: Compare distances of hiking trails.

Cooperative Group Activity

Stick It in Order

Materials: self-stick notes, pencils, paper

Procedure: Organize students into groups of three. Give each student six self-stick notes.

- Have each student in the group write a whole number on the first note, a proper fraction on the second, and a mixed number on the third. All numbers should be less than 10.
- The group then works together to order the three numbers from least to greatest. Each group then sticks the notes in order on the chalkboard.
- A volunteer from each group explains how the group arrived at the order it used. Other groups check the work on the chalkboard.
- Students repeat the process until they use up all six of their self-stick notes.

Customizing the Activity for Individual Needs

ESL Have students use *least* and *greatest* to describe relationships. For example: "Of all the shows on TV at 6 o'clock, the news is my least favorite. Of all my friends, you're the greatest!" Then have students identify the least and greatest fractions in some of the activity results.

Learning Styles Students can:

 highlight the fraction part of each mixed number.

 arrange fraction models in order.

 describe how to order the numbers by changing the different forms.

Reinforcement Activity

Provide Visual 9 *Fraction Cards* and Visual 10 *Mixed Number Cards*. Have students cut out the cards, shuffle them, and place them face down. Students select three of four cards, place the cards face up, and order them from least to greatest.

Alternative Assessment

Students can use fraction bars to order fractions, whole numbers, and mixed numbers from least to greatest.

Example: Place the following numbers in order from least to greatest:

$\frac{3}{4}, \frac{3}{2}, 1\frac{1}{8}.$

Answer:

$\frac{3}{4}$

$1\frac{1}{8}$

$\frac{3}{2}$

7.11 Problem Solving: Patterns *Student Edition pages 148–149*

Prerequisite Skill
- Comparing and ordering fractions

Lesson Objectives
- Complete fraction patterns.
- Find one number in a pattern.

Cooperative Group Activity

Pattern Puzzles

Materials: pencils, paper

Procedure: Organize students into small groups. Write the following on the chalkboard:

$\frac{1}{10}, \frac{1}{5}, \frac{3}{10}, \frac{2}{5}, \underline{}, \underline{}, \underline{}, \underline{}, \frac{9}{10}, 1$

$1, \frac{11}{12}, \frac{5}{6}, \underline{}, \frac{2}{3}, \underline{}, \underline{}, \frac{5}{12}, \underline{}, \frac{1}{4}$

$2\frac{1}{6}, 2\frac{1}{3}, 2\frac{1}{2}, \underline{}, \underline{}, \underline{}, \underline{}, \underline{}$

- Group members copy the sequences and work together to complete each one.

  Answers: The missing numbers in each sequence are:

  $\frac{1}{2}, \frac{3}{5}, \frac{7}{10}, \frac{4}{5}$

  $\frac{3}{4}, \frac{7}{12}, \frac{1}{2}, \frac{1}{3}$

  $2\frac{2}{3}, 2\frac{5}{6}, 3, 3\frac{1}{6}, 3\frac{1}{3}$

- For each sequence, a volunteer demonstrates to the class the work the group did to find the missing numbers.

Customizing the Activity for Individual Needs

ESL Explain that a *pattern* is the way a design repeats itself in an orderly arrangement. Discuss patterns on clothing or in art. Write the sequence 3, 6, 9, 12, __, __ on the board. Ask students to find the missing numbers and explain the rule for the pattern.

Learning Styles Students can:

 shade fraction pieces to show equivalency and build patterns.

 model each number in the pattern by shading fraction bars or by drawing their own models.

 say the numbers in the sequence aloud and describe the pattern in words.

Reteaching Activity

Draw the following pattern of shapes on the chalkboard.

Ask students to draw the shape that comes next.

Alternative Assessment

Students can explain how to find the missing numbers in a pattern with fractions.

Example: Explain how you would find the missing terms in the pattern. What denominator would you choose? Why? What is the pattern? What are the missing numbers in lowest terms?

$\frac{1}{8}, \frac{3}{16}, \frac{1}{4}, \underline{}, \underline{}, \frac{7}{16}, \underline{}, \frac{9}{16}$

Answers: 16 because it is the least common denominator for 4, 8, and 16. The pattern is

$\frac{2}{16}, \frac{3}{16}, \frac{4}{16}, \frac{5}{16}, \frac{6}{16}, \frac{7}{16}, \frac{8}{16}, \frac{9}{16}.$

The missing numbers are $\frac{5}{16}, \frac{3}{8},$ and $\frac{1}{2}$.

Closing the Chapter
Student Edition pages 150–151

Chapter Vocabulary

Review with students the Words to Know on page 129 of the Student Edition. Then have students quiz each other in pairs.

Have students copy and complete the Vocabulary Review questions on page 150.

For more vocabulary practice, have them complete Vocabulary 77 *Fractions and Mixed Numbers* from the Classroom Resource Binder.

Test Tips

Have pairs of students take turns modeling and explaining each test tip.

Learning Objectives

Have students review *Goals and Self-Check* from page CM6 of the Classroom Resource Binder. They can check off the goal they have reached. Note that each section of the quiz corresponds to a Learning Objective.

Group Activity

Summary: Students look up three stock prices and order them from least to greatest.

Materials: newspapers, pencils, paper

Procedure: Provide each student with a copy of the financial section of a newspaper. Direct students to choose any three stock prices and list them. They then need to order them from least to greatest.

Assessment: Use the Group Activity Rubric on page *xii* of this guide. Fill in the rubric with the additional information below. For this project, students should have:

- listed three stock prices that are fractions or mixed numbers.
- ordered three stock prices correctly from least to greatest.

RELATED MATERIALS See the unit overview page for other Globe Fearon books that can be used to enrich and extend the material in this unit.

Assessing the Chapter

Traditional Assessment

Chapter Quiz
The Chapter Quiz on pages 150–151 of the Student Edition can be used as either an open- or a closed-book test, or as homework. The quiz can be used to identify concepts in the chapter that students need to review and practice.

Chapter Tests
Use Chapter Test A Exercise 85, and Chapter Test B Exercise 86 *Fractions and Mixed Numbers* from the Classroom Resource Binder to further assess mastery of chapter concepts.

Additional Resources
Use the Resource Planner on pages 73–74 of this guide to assign additional exercises from the Classroom Resource Binder and Workbook.

Alternative Assessment

Interview
Write one fraction, one mixed number, and one whole number on a piece of paper. Ask:
- Which is the fraction?
- What is the name of this fraction?
- Draw a picture of the fraction.
- Which is the mixed number?
- What is the name of this mixed number?
- Change the mixed number to a fraction.

As-Large-As-Life Numbers
Have students list at least five places where they would expect to find, read, hear, or use fractions or mixed numbers. For each place, ask them to explain why fractions or mixed numbers are used. Some ideas students might consider are:
- financial section of a newspaper.
- cookbook.
- building construction.
- TV programs.
- things about themselves.

Planning the Chapter

Chapter 8 • Multiplying and Dividing Fractions

Chapter at a Glance

SE page | Lesson
152 | Chapter Opener and Project
154 | 8.1 Multiplying Fractions
156 | 8.2 Canceling
158 | 8.3 Multiplying Fractions and Whole Numbers
159 | Using Your Calculator: Multiplying Fractions and Whole Numbers
160 | 8.4 Multiplying Mixed Numbers
162 | 8.5 Dividing by Fractions
164 | 8.6 Dividing Fractions by Whole Numbers
165 | On-The-Job Math: Car Rental Agent
166 | 8.7 Problem Solving: Solve a Simpler Problem
168 | 8.8 Dividing Mixed Numbers
170 | 8.9 Dividing Mixed Numbers by Mixed Numbers
172 | 8.10 Problem Solving: Does the Answer Make Sense?
174 | Chapter Review and Group Activity

Learning Objectives
- Multiply fractions.
- Multiply fractions using canceling.
- Multiply fractions, whole numbers, and mixed numbers.
- Divide fractions by fractions.
- Divide whole numbers, mixed numbers, and fractions.
- Apply multiplying fractions to finding gallons of gasoline.
- Solve word problems by multiplying and dividing fractions.

Life Skills
- Alter recipe amounts.
- Compare food costs.
- Use a calculator to multiply fractions by whole numbers.
- Understand musical notes.
- Calculate overtime pay.

Communication Skills
- Read a gasoline gauge.
- Draw a picture to solve a problem.
- Write a letter.

Workplace Skills
- Estimate how much gasoline a car needs.
- Calculate the number of gallons needed to fill a fuel tank.

PREREQUISITE SKILLS
To assess mastery of prerequisite skills, use a selection of exercises from any of the program resources referenced below. The same resources can be used to provide remediation if necessary.

Skills	Program Resources for Review		
	Student Edition	Workbook	Classroom Resource Binder
Multiplying whole numbers	Lesson 4.2	Exercise 27	
Reducing fractions to lowest terms	Lesson 7.3	Exercise 56	Review 78
Changing a mixed number to an improper fraction	Lesson 7.9	Exercise 62	Practice 82
Changing an improper fraction to a mixed number	Lesson 7.8	Exercise 61	Practice 82

Diagnostic and Placement Guide

**Chapter 8
Multiplying and Dividing Fractions**

The Percent Accuracy scores are based on the number of problems in a lesson that have been answered correctly.

Resource Planner

After diagnosing your students' needs, use the correlating Program Resources for reinforcement, reteaching, or enrichment. Additional activities for customizing the lessons can be found in this guide.

KEY
Reteaching = ⤴   Reinforcement = ⬇   Enrichment = ⤴

		Percent Accuracy				Program Resources			
	Lessons	50%	65%	80%	100%	⬇ Student Edition • Extra Practice	⬇ Workbook • Exercises	Teacher's Planning Guide	Classroom Resource Binder
8.1	Multiplying Fractions	12	16	19	24	p. 423	65	⤴ p. 89	⬇ Review 88 *Multiplying Fractions with and Without Canceling* ⬇ Practice 94 *Multiplying Fractions and Mixed Numbers*
8.2	Canceling	15	20	24	30	p. 423	66	⤴ p. 90	⬇ Review 88 *Multiplying Fractions with and Without Canceling* ⬇ Practice 94 *Multiplying Fractions and Mixed Numbers*
8.3	Multiplying Fractions and Whole Numbers	10	13	16	20	p. 423	67	⤴ p. 90	⬇ Review 89 *Multiplying Fractions, Whole Numbers, and Mixed Numbers* ⬇ Practice 94 *Multiplying Fractions and Mixed Numbers*
8.4	Multiplying Mixed Numbers	12	16	19	24	p. 423	68	⤴ p. 92	⬇ Review 89 *Multiplying Fractions, Whole Numbers, and Mixed Numbers* ⬇ Practice 94 *Multiplying Fractions and Mixed Numbers*
8.5	Dividing by Fractions	12	16	19	24	p. 424	69	⤴ p. 93	Visual 8 *Fraction Bars* ⬇ Review 90 *Dividing by Fractions* ⬇ Practice 95 *Dividing Fractions and Mixed Numbers*
8.6	Dividing Fractions by Whole Numbers	12	16	19	24	p. 424	70	⤴ p. 94	⬇ Review 91 *Dividing Fractions by Whole Numbers* ⬇ Practice 95 *Dividing Fractions and Mixed Numbers*
	On-The-Job Math: Car Rental Agent	*Can be used for portfolio assessment.*						⤴ p. 94	On-The-Job Math 97 *Car Rental Agent*

Chapter 8 • 87

Lessons		Percent Accuracy				Program Resources			
		50%	65%	80%	100%	Student Edition Extra Practice	Workbook Exercises	Teacher's Planning Guide	Classroom Resource Binder
8.7	Problem Solving: Solve a Simpler Problem	1	2	2	3		71	p. 95	Review 92 Problem Solving: *Solve a Simpler Problem*
8.8	Dividing Mixed Numbers	16	21	26	32	p. 424	72	p. 96	Visual 7 *Fraction Circles* Practice 95 *Dividing Fractions and Mixed Numbers* Mixed Practice 96 *Multiplying and Dividing Fractions and Mixed Numbers*
8.9	Dividing Mixed Numbers by Mixed Numbers	18	23	29	36	p. 424	73	p. 97	Practice 95 *Dividing Fractions and Mixed Numbers* Mixed Practice 96 *Multiplying and Dividing Fractions and Mixed Numbers* Challenge 98 *Stock Market Investments*
8.10	Problem Solving: Does the Answer Make Sense?	1	2	2	3		74	p. 98	Review 93 *Problem Solving: Does the Answer Make Sense?*
Chapter 8 Review									
	Vocabulary Review	3	4	5	6			p. 99	Vocabulary 87
		(writing is worth 4 points)							
	Chapter Quiz	14	18	22	28			p. 99	Chapter Test A, B 99, 100

Customizing the Chapter

Opening the Chapter
Student Edition pages 152–153

Photo Activity
Procedure: Discuss the job of an Emergency Medical Team. Have students explain why timing is so crucial. Ask a volunteer to read the caption. Have students use reasoning to determine if the amount of time needed will be greater than or less than 15 minutes. Have students decide which of the following teams would meet the 15-minute limit:

Team	Rate of Travel	Distance Away
A	2/3 mile/minutes	10 miles
B	5/6 mile/minutes	15 miles
C	9/10 mile/minutes	9 miles

Answers: Team A, 15 min; Team C, 10 min

Words to Know
Review with students the Words to Know on page 153 of the Student Edition. Help students remember these terms by having students compare the terms' everyday meanings to their mathematical meanings.

The following words and definitions are covered in this chapter.

canceling dividing a numerator and a denominator by the same number

invert to reverse the positions of the numerator and denominator of a fraction

Recipe Project

Summary: Students choose recipes and alter the amounts of the ingredients to make one serving.

Materials: newspapers, magazines, cookbooks, paper, pencils

Procedure: Students can complete this project in class with all materials provided. Or students can work on the project for homework.

Assessment: Have students hand in the original recipes and the adjusted recipes. Grading can be based on the accuracy of the ingredient amounts in the adjusted recipes. You may wish to use the Individual Activity Rubric on page *xi* of this guide.

Learning Objectives

Review with students the Learning Objectives on page 153 of the Student Edition before starting the chapter. Students can use the list of objectives as a learning guide. Suggest that students write the objectives in a journal or use CM6 *Goals and Self-Check* in the Classroom Resource Binder.

After each lesson, have students write an example of the skill they learned under the appropriate objective. Suggest that students use these notes as a learning guide to help them study for the chapter test.

8.1 Multiplying Fractions
Student Edition pages 154–155

Prerequisite Skills
- Multiplying whole numbers
- Reducing fractions to lowest terms

Lesson Objectives
- Multiply fractions.
- Life Skill: Plan fractions of a garden plot.

Cooperative Group Activity

Fold and Multiply

Materials: paper, colored pencils

Procedure: Organize students into groups of three. Give three sheets of paper and two different-colored pencils to each group. Group members will work together to find the product of two fractions. Begin with $\frac{1}{3} \times \frac{1}{4}$.

- The first student folds a sheet of paper into 3 equal parts. Then he/she opens up the paper and uses one color to shade $\frac{1}{3}$ of the page.

- The second student folds the same sheet of paper the other way into 4 equal parts. Then he/she opens up the paper and uses another color to shade $\frac{1}{4}$ of the page.

- The third student counts the total number of parts the paper is divided into and the number of sections where the two colors overlap. Then he/she writes a multiplication sentence to show this product.
Example: $\frac{1}{3} \times \frac{1}{4} = \frac{1}{12}$.

- Students repeat the activity with fresh sheets of paper for $\frac{1}{8} \times \frac{1}{2}$ and for $\frac{2}{3} \times \frac{3}{4}$.

Customizing the Activity for Individual Needs

ESL Review different ways to imply multiplication of fractions. Say the following aloud, and have students write the corresponding multiplication problem.

Find $\frac{1}{2}$ *of* $\frac{2}{3}$.

Multiply $\frac{1}{2}$ *by* $\frac{2}{3}$.

Find $\frac{1}{2}$ *times* $\frac{2}{3}$.

Find the *product* of $\frac{1}{2}$ and $\frac{2}{3}$.

Learning Styles   Students can:

 shade squares on grid paper to model fraction multiplication.

 fold paper to model fraction multiplication and then write a rule for multiplying fractions.

 explain the steps in multiplying fractions.

Enrichment Activity

Have students multiply each of the following:

$\frac{4}{5} \times \frac{2}{4}$; $\frac{4}{5} \times \frac{1}{4}$; $\frac{4}{5} \times \frac{4}{4}$; $\frac{4}{5} \times \frac{3}{4}$

Answers: $\frac{2}{5}, \frac{1}{5}, \frac{4}{5}, \frac{3}{5}$

Have them list the products in order from least to greatest along with the corresponding multiplication problem. Students may notice that $\frac{4}{5}$ is multiplied by $\frac{2}{4}, \frac{1}{4}, \frac{4}{4}$, and $\frac{3}{4}$ respectively. Encourage them to make their own series with a similar pattern for a partner to order.

Alternative Assessment

Students can multiply fractions by modeling the products on grid paper.

Example: Show the product of $\frac{1}{2}$ and $\frac{1}{5}$ by drawing a model.

Answer: $\frac{1}{10}$; shade grid paper as shown.

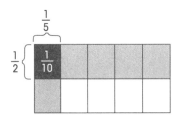

8.2 Canceling
Student Edition pages 156–157

Prerequisite Skills
- Multiplying and dividing whole numbers
- Multiplying fractions

Lesson Objectives
- Multiply fractions using canceling.
- Life Skill: Make fractional parts of recipes.

Words to Know
canceling

Cooperative Group Activity

Factor Selections

Materials: paper, pencils

Procedure: Organize students into pairs. Write the following six fractions on the chalkboard:

$\frac{5}{9}, \frac{7}{10}, \frac{15}{24}, \frac{5}{12}, \frac{8}{21}, \frac{27}{28}$

- One partner chooses two fractions from the chalkboard and writes them on paper as a multiplication problem.
- The other partner cancels common factors, if possible, and then finds the product.
- Partners switch roles and repeat the activity. They then check each other's work.

Customizing the Activity for Individual Needs

ESL Help students understand the word *canceling* by showing a canceled stamp on an envelope. Explain that the lines across the stamp prevent it from being used again. Students can circle or box pairs of numbers in the numerator and denominator that have a common factor, and then they can cross out the numbers after dividing each by the common factor.

Learning Styles Students can:

 use markers to highlight pairs of numbers in the numerator and denominator that have common factors before canceling. They then can write the division on the side.

 use number cards, which show the new numerator and denominator, to cover up the original numerator and denominator.

 say aloud the number by which the numerator and denominator can be divided as it is being done.

Reteaching Activity

Demonstrate the commutative property of multiplication. Show students that

$$\frac{3}{4} \times \frac{2}{3} = \frac{3 \times 2}{4 \times 3} = \frac{2 \times 3}{4 \times 3}$$
$$= \frac{2}{4} \times \frac{3}{3} = \frac{1}{2} \times 1 = \frac{1}{2}.$$

Encourage students to rewrite fractions so that numbers with common factors are in the same fraction. Then have students reduce the fractions to lowest terms before multiplying.

Example: $\frac{5}{8} \times \frac{2}{15}$ can be rewritten as $\frac{2}{8} \times \frac{5}{15}$. Remind students that they cannot exchange a numerator with a denominator.

Alternative Assessment

Students can explain how to cancel before multiplying two fractions.

Example: Explain how to multiply $\frac{3}{10} \times \frac{8}{9}$. Which numbers can be canceled before you multiply? Why?

Answer: Since 3 is a common factor of 3 and 9, they can be canceled by dividing each by 3. Since 2 is a common factor of 8 and 10, they can be canceled by dividing each by 2.

8.3 Multiplying Fractions and Whole Numbers
Student Edition page 158

Prerequisite Skills
- Multiplying fractions
- Reducing fractions to lowest terms
- Converting whole numbers and improper fractions

Lesson Objective
- Multiply fractions and whole numbers.

Cooperative Group Activity

Cube Toss
Materials: number cubes, paper, pencils

Procedure: Organize students into groups of three. Give three number cubes to each group.

- Students in each group toss the cubes. They choose one number for the whole number. They use the other two numbers to form a fraction. This fraction may be either proper or improper.
- Group members work together to multiply their fractions and whole numbers. They cancel before multiplying when possible.
- Students write answers that are improper fractions as mixed numbers in lowest terms.
- Each group presents its results to the class.

Customizing the Activity for Individual Needs

ESL Help students differentiate among fractional amounts that sound alike and may be easily confused, such as $\frac{2}{3}$, $2\frac{1}{3}$, and $2 \times \frac{1}{3}$. Ask students to model and describe each amount.

Learning Styles Students can:

 highlight the whole number. Then they can change it to a fraction to multiply.

 use two-color counters to model and multiply fractions and whole numbers.

 read the numbers to be multiplied to themselves while solving the problem.

Enrichment Activity

Have each student in a pair draw the outline below on paper. One student randomly draws three digits from a set of number cards (see Visual 4 *Whole Numbers to 100*). This student places the digits in the outline anywhere he or she wishes and finds the product. The other student rearranges the same digits in his or her outline and finds the product. The student who finds the greatest product earns 1 point. Continue until 10 problems have been solved. The student with the greatest number of points wins.

Alternative Assessment

Students can multiply a fraction and a whole number using two-color counters.

Example: Find $\frac{3}{4} \times 8$ using counters.

Answer: 6; arrange 8 counters of one color into 4 equal groups and turn over 3 of the 4 as shown.

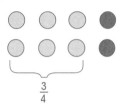

USING YOUR CALCULATOR
Multiplying Fractions and Whole Numbers
Student Edition page 159

Lesson Objective
- Life Skill: Use a calculator to multiply a whole number by a fraction.

Activities

Another Way
Materials: calculators

Procedure: Students can divide first and then multiply when multiplying fractions and whole numbers. Show students how to multiply 384 by $\frac{5}{6}$ by following these steps:

Step 1 Divide the whole number by the denominator of the fraction.

PRESS 3 8 4 ÷ 6 = | 64 |

Step 2 Leave the quotient on the screen. Multiply the quotient by the numerator of the fraction.

PRESS × 5 = | 320 |

Step 3 Write the number sentence.

$384 \times \frac{5}{6} = 320$

Variation: Students can first divide the numerator by the denominator and then multiply by the whole number.

PRESS 5 ÷ 6 × 3 8 4 = | 320 |

(Continued on p. 92)

(Continued from p. 91)
Have students multiply the following:

$\frac{3}{4} \times 112$ $290 \times \frac{21}{5}$

Answers: 84; 1,218

Calculator or Paper and Pencil?

Materials: calculators, paper, pencils

Procedure: Have students work in pairs. Provide students with multiplication problems where the denominator of the fraction divides the whole number evenly. For example:

$81 \times \frac{4}{9}$; $\frac{7}{10} \times 120$; $\frac{5}{4} \times 28$; $22 \times \frac{3}{2}$; $100 \times \frac{9}{5}$;

$175 \times \frac{2}{7}$;

$\frac{8}{25} \times 325$; and $\frac{9}{8} \times 72$

- One player uses a calculator to find the products. The other player uses multiplication and canceling on paper to find the products.
- If the products match, the player using paper scores 1 point. If they do not match, someone is incorrect. Check each other's work to determine who got the first correct answer. This player gets the point.
- Have players alternate roles and continue playing until one player reaches 8 points.

8.4 Multiplying Mixed Numbers
Student Edition pages 160–161

Prerequisite Skills
- Converting mixed numbers and improper fractions
- Multiplying fractions
- Dividing by whole numbers

Lesson Objectives
- Multiply mixed numbers.
- Life Skill: Compare food costs.

Cooperative Group Activity

Team Mixed Number Multiply

Materials: mixed numbers or Visual 10 *Mixed Number Cards*; index cards; paper; pencils

Procedure: Organize students into teams of four. Give each group a stack of index cards.

Pick two mixed numbers, and write them on the chalkboard.

- Two members of each group write the mixed numbers, each on a separate index card. The cards are passed to the other two team members.
- These two team members convert the mixed numbers to improper fractions, and they then write the improper fractions on the backs of the respective index cards.
- The whole team then works together to multiply the improper fractions. The product is written on a third card. On the back, the product is written in mixed number form.
- When the team is ready, the members raise their hands. Any team member may be called to answer. A correct answer scores a point.

Customizing the Activity for Individual Needs

ESL Review the difference between the words *improper fraction* and a *mixed number*. Write $\frac{18}{7}$ and $2\frac{4}{7}$ on the chalkboard. Discuss why $\frac{18}{7}$ is an *improper* fraction and $2\frac{4}{7}$ is a *mixed* number.

Learning Styles Students can:

 write each mixed number in a different color. Then they can write the corresponding improper fractions in those colors.

 use number cards to model mixed numbers, improper fractions, and products.

 say aloud the steps to change mixed numbers to improper fractions; to multiply fractions; and to change a product to a mixed number.

Enrichment Activity

Have students make a display of unit prices for items that can be found in a grocery store. The unit prices should be in dollar amounts.

Cheeses	Prices
Cheddar	$3/pound
American	$2/pound
Swiss	$4/pound

Using the display, have students make up their own problems for classmates to solve. Each problem should involve fractional amounts and be accompanied by a worked-out solution. Students' work can be posted on a bulletin board or made into booklets.

Alternative Assessment

Students can explain the steps for multiplying mixed numbers.

Example: Explain how to multiply $1\frac{1}{2} \times 2\frac{4}{5}$.

Answer: Explanations should include changing mixed numbers to improper fractions, multiplying the fractions, and writing the product as a mixed number in lowest terms.

8.5 Dividing by Fractions
Student Edition pages 162–163

Prerequisite Skills
- Multiplying fractions
- Reducing fractions to lowest terms

Lesson Objectives
- Divide whole numbers and fractions by fractions.
- Life Skill: Read the value of musical notes.

Words to Know
invert

Cooperative Group Activity

Fraction Shuffle

Materials: two sets of fraction cards for $\frac{1}{2}$, $\frac{1}{3}$, $\frac{2}{3}$, $\frac{1}{4}$, $\frac{3}{4}$, and $\frac{5}{6}$ or Visual 9 *Fraction Cards*; ÷ and × symbol cards; pencils; paper

Procedure: Organize students into groups of three. Give two sets of cards to each group.
- Each group shuffles the cards and places them face down on a desk.
- One member picks a card and places it face up as the fraction. This will be the first number in the problem. The ÷ symbol is placed next to it.
- A second member picks another card, writes the inverted fraction on the back, and then places it face up as the second number in the problem.
- The third member switches the operation symbols, flips over the card to show the *inverted* number, and finds the quotient.
- Students put these cards aside. They switch roles and repeat the activity until all group members have had a chance to find the quotient.

Customizing the Activity for Individual Needs
ESL Help students understand that *invert* means to *turn upside down* or to *change the position of*. Demonstrate by inverting a cup or other object. Compare it with the change in position of the numerator and denominator when a fraction is inverted. Have students draw arrows to show that the numerator becomes the denominator and the denominator becomes the numerator.

Learning Styles Students can:

 use a marker to highlight the fraction to be inverted, the inverted fraction, and the change from division to multiplication.

 cut out fraction strips to model division of a whole number by a fraction.

 say aloud the fraction to be inverted and name the inverted fraction.

Enrichment Activity

Students can make musical note puzzles to find out how many smaller notes make up a set of larger notes. Have students put the puzzles on construction paper for others to solve, using the key below. Students can write the division problems in "note code" and the number translation on the back.

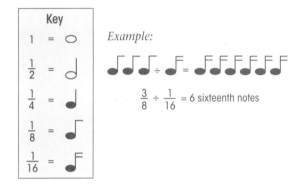

Alternative Assessment

Students can explain how to divide by a fraction.

Example: Divide $\frac{1}{5}$ by $\frac{3}{10}$. Which fraction would you change? How would you change it? What else would you change?

Answer: $\frac{2}{3}$; invert the second fraction; change it to $\frac{10}{3}$; change the division problem to a multiplication problem.

8.6 Dividing Fractions by Whole Numbers
Student Edition page 164

Prerequisite Skills
- Multiplying fractions
- Dividing by fractions

Lesson Objective
- Divide a fraction by a whole number.

Cooperative Group Activity

Divide the Fraction
Materials: digit cards for fractions and whole numbers or Visual 4 *Whole Numbers to 100*, and Visual 9 *Fraction Cards*; multiplication and division symbol cards; paper; pencils

Procedure: Organize students into groups of three. Give each group a set of whole number cards and a set of fraction cards.
- Each group shuffles the whole number and fraction cards separately and then places them face down in two piles.
- One member picks a fraction card and places it face up in front of the division symbol.
- The second member picks a whole number card, writes the inverted form on the back, and then places it face up after the division symbol.
- The third member changes the division to multiplication and *inverts* the whole number.
- Members work together to find the quotient.
- Students replace the cards and shuffle again. They switch roles and repeat until all members have had a chance to find the quotient.

Customizing the Activity for Individual Needs
ESL To help students understand the word *invert*, groups should brainstorm other words that mean *invert*, such as *flip over, turn upside down,* or *switch around*. Ask students to gesture as they say the words.

Learning Styles Students can:

 highlight the whole number that divides the fraction.

 use number cards to model fractions, whole numbers, and the quotients.

 say steps for dividing fractions by whole numbers aloud when solving problems.

Reteaching Activity
Have students use fraction circles or Visual 7 *Fraction Circles* to cut out fraction wedges and then cut or fold again to show division by a whole number.

Have students find $\frac{1}{4} \div 2$ by cutting out a $\frac{1}{4}$ wedge and then cutting it in half as shown below. They can then match up the resulting wedge to a *whole* circle to identify its size.
Repeat for $\frac{1}{8} \times 2$ and $\frac{1}{2} \div 3$.

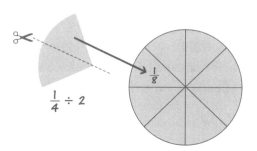

Alternative Assessment
Students can divide a fraction by a whole number by filling in missing steps correctly.

Example: Complete the steps to the following division problem:

$$\frac{2}{3} \div 4$$

$$\frac{2}{3} \;?\; \frac{\square}{\square} = \frac{\square}{\square} = \frac{\square}{\square}$$

Answer: $\frac{2}{3} \div 4 = \frac{2}{3} \times \frac{1}{4} = \frac{2}{12} = \frac{1}{6}$

ON-THE-JOB MATH
Car Rental Agent
Student Edition page 165

Lesson Objectives
- Apply multiplying fractions to finding gallons of gasoline.
- Life Skill: Read a gasoline gauge.
- Communication Skill: Write a letter to request a car brochure.
- Workplace Skill: Calculate the number of gallons needed to fill a fuel tank.

Activities

Tools of the Trade

Materials: pencil, paper

Procedure: Point out to students that the fuel gauge in a car tells how much gasoline is left in the fuel tank. This is a fractional part of a full tank. The actual amount depends on the size of the tank. One-fourth of a 20-gallon tank will be less than one-fourth of a 24-gallon tank.

20-gallon tank
$\frac{1}{4} \times 20 = 5$ gallons

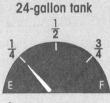
24-gallon tank
$\frac{1}{4} \times 24 = 6$ gallons

- Have students complete the following table to find the number of gallons in each tank for each reading. Students can then compare the results.

Tank Holds	Gauge at $\frac{1}{4}$	$\frac{1}{2}$	$\frac{3}{4}$	Full
16 gallons	4 gal		12 gal	16 gal
20 gallons				
24 gallons				
32 gallons				

Answers: 8 gal; 5 gal, 10 gal, 15 gal, 20 gal; 6 gal, 12 gal, 18 gal, 24 gal; 8 gal, 16 gal, 24 gal, 32 gal

- Students can use the information in the table to create puzzles for others to solve.

  Example: My fuel tank is $\frac{1}{2}$ full. There are 12 gallons left in the tank. How many gallons are in a full tank?

  Answer: 24 gallons

Write a Business Letter

Materials: paper and pen or computer word processor

Procedure: Have students write letters to two different automobile manufacturers or new car dealerships to request brochures for the models of automobiles they sell. Students should ask for information about
- the size of the fuel tank in each model.
- the grade of gasoline that is recommended for each model.
- the number of miles per gallon achieved by each model.

Act It Out

Materials: none

Procedure: Have students pretend they are car rental agents. A customer returns a car.

The gasoline gauge reads between $\frac{1}{4}$ and $\frac{1}{2}$ of a tank. The car rental agent wants the customer to pay for $\frac{3}{4}$ of a tank of gasoline. The customer feels that this is too much since there is more than $\frac{1}{4}$ tank of gasoline in the tank. Have students suggest a way to determine how much gasoline the customer should pay for that is fair to both agent and customer.

Practice

Have students complete On-The-Job Math 97 *Car Rental Agent* from the Classroom Resource Binder.

Internet Connection

Students can locate the Web sites of various car rental agencies to look for links about employment opportunities.

Avis
    http://www.avis.com
National Car Rental
    http://www.nationalcar.com

8.7 Problem Solving: Solve a Simpler Problem
Student Edition pages 166–167

Prerequisite Skills
- Multiplying fractions, whole numbers, and mixed numbers
- Dividing fractions and whole numbers
- Reducing fractions to lowest terms

Lesson Objectives
- Use whole numbers to determine the operation to use in a word problem involving fractions.
- Solve word problems by multiplying or dividing fractions.
- Draw pictures to solve problems.

Cooperative Group Activity

Simply Cross It Out

Materials: index cards, paper, pencils, red pens

Procedure: Organize students into pairs. Give one index card and a red pen to each student. Encourage students to use their textbook for ideas and examples.

- Each student writes a word problem on the index card that involves at least one fraction or mixed number.
- Pairs exchange problems. After reading the problem, the student uses a red pen to cross out any fraction or mixed number. The student then writes a whole number that makes sense above it. Finally, each student writes a plan to solve his or her problem.
- The original author of the word problem checks the plan and makes suggestions if needed.
- Each student then follows the plan to solve the problem using the original fraction or mixed number.
- Partners check each other's work.

Customizing the Activity for Individual Needs

ESL To help students understand the *Plan* step in problem solving, discuss the use of an outline in writing and a blueprint in construction.

Learning Styles Students can:

 use self-stick notes to cover difficult numbers in a problem and replace them with simpler numbers. Then they can write the plan.

 use real-life materials to model a problem.

 read aloud problems and explain another way to solve them.

Enrichment Activity

Have partners use various fraction models, such as Visual 6 *Fraction Sets,* Visual 7 *Fraction Circles,* and Visual 8, *Fraction Bars,* to make up problems based on pictures they create. For example, one student makes a picture like the one shown. The other writes a problem for it and a solution.

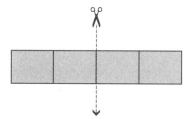

I have a string 1 inch long. I cut it in half. How long is each piece?

Answer: $1 \div 2 = \frac{1}{2}$ inch

Alternative Assessment

Students can explain how to solve a word problem involving fractions by using simpler whole numbers.

Example: Explain how to solve the problem below. What can you change to make the problem simpler? Will you multiply or divide? Why?

> To make 1 cup of cocoa, Sheila uses $\frac{3}{4}$ of a cup of milk. How much milk should she use to make 8 cups of cocoa?

Answer: Use a whole number instead of $\frac{3}{4}$ for the amount of milk. Multiply since she is making more than 1 cup. 6 cups of milk

8.8 Dividing Mixed Numbers
Student Edition pages 168–169

Prerequisite Skills
- Dividing fractions
- Converting mixed numbers and improper fractions

Lesson Objectives
- Divide fractions by mixed numbers.
- Divide mixed numbers by fractions.
- Life Skill: Read a time chart.

Cooperative Group Activity

Picture Division

Materials: whole circles and those divided into 2, 3, 4, 5, 6, 8, 10, and 12 equal sections or Visual 7 *Fraction Circles;* scissors; paper; pencils

Procedure: Organize the class into groups of three. Give four copies of Visual 7 *Fraction Circles* to each group. Ask students to model the division problem $2\frac{1}{2} \div \frac{1}{4}$. Remind them that $2\frac{1}{2} \div \frac{1}{4}$ means this:

"How many fourths are in $2\frac{1}{2}$?"

- One student cuts out the circles that will be needed to find the quotient, producing two whole circles, one half-circle, and one quarter-circle.
- Another student counts the number of quarter circles that fit on $2\frac{1}{2}$ circles and writes the quotient.

- A third student shows how to change the division problem into a multiplication problem and then finds the product. Students compare this product with the quotient found using the manipulatives.
- Students repeat the activity for $2\frac{1}{4} \div \frac{1}{8}$, $1\frac{3}{4} \div \frac{1}{8}$, and $2\frac{1}{2} \div \frac{1}{6}$.

Customizing the Activity for Individual Needs

ESL To help students remember the word *division*, review words and phrases such as *divided by, goes into, how many...are in*. Encourage pairs to take turns saying the same division problems using different wordings.

Learning Styles Students can:

 draw models to show the division problem and the quotient.

 cut out and rearrange fraction pieces into model problems and then find quotients.

 orally explain how to find quotients while counting the number of pieces that fit on the given number of circles.

Enrichment Activity

Have students create a poster of the steps in the lesson to show how to divide a fraction by a mixed number or a mixed number by a fraction. Display posters around the room.

Alternative Assessment

Students can show their understanding of division with mixed numbers by filling in the missing numbers in a calculation.

Example: Complete the following problem:
$\frac{1}{4} \div 2\frac{1}{2} = \frac{1}{4} \div ? = \frac{1}{4} \times ? = ?$

Answer: $\frac{1}{4} \div 2\frac{1}{2} = \frac{1}{4} \div \frac{5}{2} = \frac{1}{4} \times \frac{2}{5} = \frac{1}{10}$

8.9 Dividing Mixed Numbers by Mixed Numbers
Student Edition pages 170–171

Prerequisite Skills
- Dividing fractions
- Converting mixed numbers and improper fractions

Lesson Objectives
- Divide a mixed number by a mixed number.
- Life Skill: Calculate overtime pay.

Cooperative Group Activity

Division Mixup

Materials: index cards, pencils

Procedure: Organize students into pairs. Give each pair of students at least one index card.

- Each student makes up a mixed number. Working together, pairs find the quotient of the mixed numbers.
- Students write their mixed number division problem on one side of the index card. They write the quotient on the other side.
- Pairs exchange index cards with other groups who solve the problems.
- Pairs check that their quotients are correct by looking at the backs of the cards.
- *Extension*: Students can rearrange the mixed numbers in a division problem to find the larger number and the smaller number.
  Example: $2\frac{1}{2} \div 1\frac{2}{3} = \frac{3}{2}$ $1\frac{2}{3} \div 2\frac{1}{2} = \frac{2}{3}$

Customizing the Activity for Individual Needs

ESL To help students understand the words *mixed number, improper fraction, division sign, multiplication sign, invert, cancel, simplify,* and *quotient*, label those parts on a worked out division problem.

Learning Styles Students can:

 highlight the second number in the division problem to remind each student to invert and multiply.

 arrange digit cards to show the division of mixed numbers.

 recite the steps to themselves while completing the problem.

Enrichment Activity

Have small groups make up their own weekly time cards. Tell them to include overtime hours (more than 40 hours), and they can choose their own pay rates. Overtime pay is time-and-a-half. Then students switch with a partner to calculate the pay.

Chapter 8 • 97

Alternative Assessment

Students can show their understanding of dividing a mixed number by a mixed number by filling in the missing numbers in a calculation.

Example: Complete the following problem:
$2\frac{1}{3} \div 3\frac{2}{3} = \frac{?}{3} \div \frac{?}{3} = \frac{?}{3} \times \frac{3}{?} = ?$

Answer: $\frac{7}{3} \div \frac{11}{3} = \frac{7}{3} \times \frac{3}{11} = \frac{7}{11}$

8.10 Problem Solving: Does the Answer Make Sense?
Student Edition pages 172–173

Prerequisite Skills
- Multiplying and dividing fractions, whole numbers, and mixed numbers

Lesson Objective
- Solve fraction word problems.

Cooperative Group Activity

No Nonsense Ad-libs

Materials: pencils, index cards

Procedure: Organize students into small groups. Have students make up *ad-lib* word problems. Write the following setups on the chalkboard, or provide copies on paper.

_____ listened to _____ for _____ minutes.
Name 1   Any band   A. Mixed number

_____ listened for _____ as long. How long did _____
Name 2   B. Fraction   Name 2

listen to the band? _____ listened for _____ minutes.
                          Name 2   C. Number

To solve: A × B = C; C ÷ B = A; C ÷ A = B

- One group member copies a blank ad-lib problem onto an index card. To fill in the blanks, he or she interviews another member asking for specific words or numbers for each blank.
- The group discusses the problem and decides if the number in C makes sense.
- A different student then uses the formula to solve for C and checks. If it is wrong, the student fixes the problem.
- Now the group blocks out one of the three numbers and writes it on the back as the solution. The card now goes to another group for it to solve for the missing number.

Extension: Students can make up their own ad-lib blanks for others to complete.

Customizing the Activity for Individual Needs

ESL To help students determine whether the answer will be a quantity or a measure, have them look for the key words in the word problem question. *How many* asks for a quantity, such as 20 books. *How long* asks for a measure, such as length or time. As students write the numbers for the problem, have them label what each number represents.

Learning Styles Students can:

 draw pictures to illustrate problems and answers to see if they make sense.

 use real-life pictures/objects to model problems and answers to see if they make sense.

 say the solution aloud in a complete sentence to decide if it makes sense.

Enrichment Activity

Have students turn written-response word problems into multiple-choice problems. They may use problems from their textbook or those created during the Cooperative Group Activity. Ask them to think about each incorrect choice and why it is the wrong answer. Encourage them to make up choices that will *fool* their classmates. Have classmates try to choose the answers that make sense.

Alternative Assessment

Given an answer to a word problem, students can decide whether it makes sense.

Example: Anton skates $1\frac{3}{4}$ hours each day. He skated for 8 days. How long did he skate? Answer: 14 hours. Does the answer make sense? Explain.

Answer: Yes. $8 \times 1\frac{3}{4}$ should be greater than 8×1 and less than 8×2.

Closing the Chapter
Student Edition pages 174–175

Chapter Vocabulary
Review with students the Words to Know on page 153 of the Student Edition. Then have students give examples that illustrate the meaning of each word.

Have students copy and complete the Vocabulary Review questions on page 174 of the Student Edition.

For more vocabulary practice, have them complete Vocabulary 87 *Multiplying and Dividing Fractions* from the Classroom Resource Binder.

Test Tips
Have student pairs take turns modeling and explaining each of the test tips.

Learning Objectives
Have students review CM6 *Goals and Self-Check*. They can check off the goal they have reached. Note that each section of the quiz corresponds to a Learning Objective.

Group Activity
Summary: Students alter amounts of ingredients and numbers of servings in three recipes.

Materials: three recipes students gathered for the chapter Recipe Project or newspapers, magazines, or cookbooks; pencils, paper

Procedure: Have students show the original recipe and the altered recipe. The work that was done to calculate the amount of each ingredient in the altered recipe can be stapled to the recipe. Encourage students to label their work carefully.

Assessment: Use the Group Activity Rubric on page *xii* of this guide. Fill in the rubric with the additional information below. For this project, students should have:
- chosen three recipes and correctly altered the amounts for all ingredients in all three recipes.
- given the correct new number of servings for all three recipes.

RELATED MATERIALS See the unit overview page for other Globe Fearon books that can be used to enrich and extend the material in this unit.

Assessing the Chapter

Traditional Assessment
Chapter Quiz
The Chapter Quiz on pages 174–175 of the Student Edition can be used as an open-book test, a closed-book test, or as homework. The quiz can be used to identify concepts in the chapter that students need to review and practice.

Chapter Tests
Use Chapter Test A Exercise 99 and Chapter Test B Exercise 100 *Multiplying and Dividing Fractions* from the Classroom Resource Binder to further assess mastery of chapter concepts.

Additional Resources
Use the Resource Planner on pages 87–88 of this guide to assign additional exercises from the Classroom Resource Binder and Workbook.

Alternative Assessment
Demonstration
Have students use fraction circles to find:
- the product of $3\frac{1}{2} \times 2$.
- the product of $2\frac{3}{4} \div \frac{1}{4}$.

Comparing Numbers
Display the following problems:
$3\frac{1}{2} \times \frac{2}{3}$ and $3\frac{1}{2} \div \frac{2}{3}$. Ask:
- How are the problems alike or different?
- Will the answers be the same? Explain.
- In which problem would you invert a number before solving? Which number do you invert?
- Find the product and the quotient.

Planning the Chapter

Chapter 9 • Adding and Subtracting Fractions

Chapter at a Glance

SE page	Lesson	
176		*Chapter Opener and Project*
178	9.1	Adding and Subtracting Like Fractions
180	9.2	Adding Like Mixed Numbers
182	9.3	Subtracting Like Mixed Numbers
184	9.4	Subtracting from a Whole Number
186	9.5	Adding Unlike Fractions
187		*Math in Your Life: Pay Day*
188	9.6	Subtracting Unlike Fractions
189		*Using Your Calculator: Finding Common Denominators*
190	9.7	Adding Unlike Mixed Numbers
192	9.8	Subtracting Unlike Mixed Numbers
194	9.9	Problem Solving: Multi-Part Problems
196		*Chapter Review and Group Activity*
198		*Unit Review*

Learning Objectives
- Add and subtract like fractions.
- Add and subtract like mixed numbers.
- Subtract mixed numbers and fractions from whole numbers.
- Add and subtract unlike fractions.
- Add and subtract unlike mixed numbers.
- Solve multi-part fraction problems.
- Apply adding and subtracting fractions to find work hours.

Life Skills
- Track the performance of stocks.
- Use a calculator to find common denominators.
- Calculate time on the job.
- Calculate a week's pay.
- Use a map to find the shortest route.

Communication Skills
- Keep a record of the time spent in after-school activities.
- Interpret tables, charts, and maps.
- Explain why a common denominator is needed to add two unlike fractions.
- Write a letter.
- Report the activity of several stocks using the newspaper as a resource.

PREREQUISITE SKILLS

To assess mastery of prerequisite skills, use a selection of exercises from any of the program resources referenced below. The same resources can be used to provide remediation if necessary.

	Program Resources for Review		
Skills	Student Edition	Workbook	Classroom Resource Binder
Reducing fractions to lowest terms	Lesson 7.3	Exercise 56	Review 78
Changing fractions to mixed numbers	Lesson 7.8	Exercise 61	Practice 82
Changing whole and mixed numbers to fractions	Lesson 7.9	Exercise 62	Practice 82
Finding common denominators	Lesson 7.5	Exercise 58	Review 79
Multiplying fractions and mixed numbers	Lessons 8.1–8.4	Exercises 65–68	Review 88, 89 Practice 94 Mixed Practice 96

Diagnostic and Placement Guide

**Chapter 9
Adding and
Subtracting Fractions**

The Percent Accuracy scores are based on the number of problems in a lesson that have been answered correctly.

Resource Planner

After diagnosing your students' needs, use the correlating Program Resources for reinforcement, reteaching, or enrichment. Additional activities for customizing lessons can be found in this guide.

KEY
Reteaching = ↶   Reinforcement = ↓   Enrichment = ↷

	Lessons	Percent Accuracy				Program Resources			
		50%	65%	80%	100%	↓ Student Edition Extra Practice	↓ Workbook Exercises	Teacher's Planning Guide	Classroom Resource Binder
9.1	Adding and Subtracting Like Fractions	10	13	16	20	p. 425	75	↓ p. 103	Visual 7 *Fraction Circles* Visual 8 *Fraction Bars* ↓ Review 102 *Adding and Subtracting Like Fractions* ↓ Practice 109 *Adding Like Fractions and Like Mixed Numbers* ↓ Practice 110 *Subtracting Like Fractions and Like Mixed Numbers*
9.2	Adding Like Mixed Numbers	10	13	16	20	p. 425 p. 428	76	↓ p. 104	Visual 7 *Fraction Circles* ↓ Review 103 *Adding Like Mixed Numbers* ↓ Practice 109 *Adding Like Fractions and Like Mixed Numbers*
9.3	Subtracting Like Mixed Numbers	9	12	14	18	p. 425	77	↶ p. 105	Visual 7 *Fraction Circles* Visual 8 *Fraction Bars* ↓ Review 104 *Subtracting Like Mixed Numbers* ↓ Practice 110 *Subtracting Like Fractions and Like Mixed Numbers*
9.4	Subtracting from a Whole Number	8	10	13	16	p. 426 p. 428	78	↶ p. 106	Visual 7 *Fraction Circles* ↓ Review 105 *Subtracting from a Whole Number*
9.5	Adding Unlike Fractions	8	10	13	16	p. 426 p. 428	79	↷ p. 107	Visual 8 *Fraction Bars* ↓ Review 106 *Adding and Subtracting Unlike Fractions* ↓ Practice 111 *Adding Unlike Fractions and Unlike Mixed Numbers*
	Math in Your Life: Pay Day	*Can be used for portfolio assessment.*						↷ p. 108	Math in Your Life 115 *Pay Day*

Chapter 9

Lessons		Percent Accuracy				Program Resources			
		50%	65%	80%	100%	Student Edition Extra Practice	Workbook Exercises	Teacher's Planning Guide	Classroom Resource Binder
9.6	Subtracting Unlike Fractions	8	10	13	16	p. 426	80	p. 108	Visual 7 *Fraction Circles* Visual 8 *Fraction Bars* Review 106 *Adding and Subtracting Unlike Fractions* Practice 112 *Subtracting Unlike Fractions and Unlike Mixed Numbers*
9.7	Adding Unlike Mixed Numbers	12	16	19	24	p. 427 p. 428	81	p. 110	Review 107 *Adding Unlike Mixed Numbers* Practice 111 *Adding Unlike Fractions and Unlike Mixed Numbers*
9.8	Subtracting Unlike Mixed Numbers	9	12	14	18	p. 427 p. 428	82	p. 110	Review 108 *Subtracting Unlike Mixed Numbers* Practice 112 *Subtracting Unlike Fractions and Unlike Mixed Numbers* Mixed Practice 114 *Adding and Subtracting Fractions and Mixed Numbers* Challenge 116 *Choosing Delivery Routes*
9.9	Problem Solving: Multi-Part Problems	1	2	2	3	p. 428	83	p. 111	Visual 22 *Problem Solving Steps* Practice 113 *Problem Solving: Multi-Part Problems*
Chapter 9 Review									
	Vocabulary Review	4	5.5	6.5	8			p. 112	Vocabulary 101
		(writing worth 4 points)							
	Chapter Quiz	10	13	16	20			p. 113	Chapter Tests A, B 117, 118
Unit 2 Review		5	6.5	8	10			p. 113	Unit Test p. T9
		(critical thinking worth 4 points)							

Customizing the Chapter

Opening the Chapter
Student Edition pages 176–177

Photo Activity
Procedure: Lead a class discussion about how fractions are used in measuring inches, feet, yards, miles, and other customary units. Have students study the photograph while a volunteer reads the caption. Then make a class list of professions (in addition to drafting) in which taking careful measurements plays a key role. Examples of such professions include engineering, carpentry, architecture, and clothing design.

Words to Know
Review with students the Words to Know on page 177 of the Student Edition. Help students remember these words by making a two-column chart with the headings *Like Fractions* and *Unlike Fractions*. Have students write examples for each column.

The following words and definitions are covered in this chapter.

like fractions fractions that have the same denominator

unlike fractions fractions that have different denominators

After-School Activities Project
Summary: Students keep a record of time spent in after-school activities each day for a week using fractions of an hour, and then they find the total.

Materials: clock or watch, pencils, paper, blank three-column chart or Visual 16 *Blank Table*

Procedure: Remind students how to change minutes to a fraction of an hour by first using a denominator of 60 and then reducing the fraction.

Examples: 15 minutes = $\frac{15}{60}$ = $\frac{1}{4}$ hour;

30 minutes = $\frac{30}{60}$ = $\frac{1}{2}$ hour;

10 minutes = $\frac{10}{60}$ = $\frac{1}{6}$ hour.

Give each student a copy of Visual 16 *Blank Table*, or have them make a three-column chart. Have students use the chart to record their data.

Students can complete the first part of this project over the course of 1 week. They can find totals for the project during math class or as homework after completing Lesson 9.2. You may wish to limit the data collection to the time during which you teach the chapter.

Assessment: Use the Individual Activity Rubric on page *xi* of this guide. Fill in the rubric with the additional information below. For this project, students should have:

- kept a list of time spent on after-school activities.
- found the total hours.

Learning Objectives
Review the Learning Objectives on page 177 of the Student Edition before starting the chapter. Students can use the list as a learning guide. Suggest they write the objectives in a journal or use CM6, *Goals and Self-Check*, in the Classroom Resource Binder.

After each lesson, have students write an example of the skill they learned under the appropriate objective. Suggest that students use these notes as a learning guide to help them study for the chapter test.

9.1 Adding and Subtracting Like Fractions
Student Edition pages 178–179

Prerequisite Skills
- Reducing fractions to lowest terms
- Changing fractions to mixed numbers

Lesson Objectives
- Add and subtract like fractions.
- Life Skill: Compare sizes of items in ounces.

Words to Know
like fractions

Cooperative Group Activity

Making Fraction Bar Problems
Materials: fraction strips for wholes, halves, thirds, fourths, fifths, sixths, eighths, tenths, and twelfths or Visual 8 *Fraction Bars*; markers or colored pencils; scissors

Procedure: Organize students into small groups. Give each student two copies of Visual 8 or fraction strips. Write $\frac{1}{6} + \frac{3}{6}$ on the board. Then write $\frac{7}{12} - \frac{5}{12}$ on the board.

- Using one copy of Visual 8, each student cuts out the sixths fraction bar. One student colors the bar to show $\frac{1}{6}$. Another colors $\frac{3}{6}$. A third arranges the bars on a sheet of paper and writes the problem $\frac{1}{6} + \frac{3}{6} = \frac{4}{6}$. A volunteer reduces the sum to lowest terms: $\frac{2}{3}$.
- Group members switch roles and model addition problems for each of the remaining bars on the visual (excluding the whole), starting with halves, then thirds, fourths, fifths, eighths, tenths, and twelfths.
- Using the other copy of Visual 8, groups find differences starting with twelfths: $\frac{7}{12} - \frac{5}{12} = \frac{2}{12}$ or $\frac{1}{6}$.
- Groups continue to model subtraction problems for each fraction on Visual 8.

Customizing the Activity for Individual Needs

ESL To be sure that students understand the words *like fractions*, have them give examples of like fractions and record them in their notebooks.

Learning Styles Students can:

 use fraction bars to add and subtract like fractions.

 cut out and rearrange circular fraction models from Visual 7 *Fraction Circles* to find sums and differences.

 record addition and subtraction problems as other students read them aloud.

Enrichment Activity

Have each student make a chart of bottles of juice that are for sale in three sizes:

small $\frac{1}{9}$ oz; medium $\frac{4}{9}$ oz; and large $\frac{8}{9}$ oz.

Students can create names for the juice and write addition and subtraction word problems. Encourage students to make other sizes using a common denominator. Post the charts and word problems for others to solve.

Alternative Assessment

Students can model addition and subtraction of like fractions using fraction bars.

Examples: Use fraction bars to show why $\frac{5}{8} + \frac{7}{8} = 1\frac{1}{2}$ and $\frac{9}{10} - \frac{1}{10} = \frac{4}{5}$.

Answers: Using fraction bars for addition,

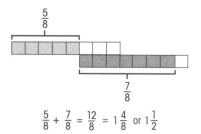

$\frac{5}{8} + \frac{7}{8} = \frac{12}{8} = 1\frac{4}{8}$ or $1\frac{1}{2}$

Using a fraction bar for subtraction,

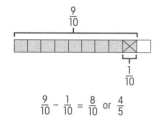

$\frac{9}{10} - \frac{1}{10} = \frac{8}{10}$ or $\frac{4}{5}$

9.2 Adding Like Mixed Numbers
Student Edition pages 180–181

Prerequisite Skills
- Reducing fractions to lowest terms
- Changing fractions to mixed numbers

Lesson Objectives
- Add like mixed numbers.
- Life Skill: Compare hours of time in a table.

Cooperative Group Activity

Adding Fraction Circles

Materials: circles divided into wholes, halves, thirds, fourths, fifths, sixths, eighths, tenths, and twelfths or Visual 7 *Fraction Circles*; markers or colored pencils; scissors

Procedure: Have students work in pairs. Give each student five copies of Visual 7. Write the following problems on the chalkboard:

$2\frac{2}{3} + 1\frac{1}{3}$ $1\frac{3}{4} + 1\frac{3}{4}$

$1\frac{3}{5} + 1\frac{1}{5}$ $2\frac{3}{4} + 1\frac{3}{4}$

Model a problem with the class: $1\frac{5}{6} + 1\frac{4}{6}$.

- One student shades sixths models to show $1\frac{5}{6}$, while the other shades $1\frac{4}{6}$. Then the students cut out the circles and arrange them on a sheet of paper to show the addition problem. The first student writes the problem under the models.

104 • Chapter 9

- To solve, they count whole shaded circles. Then they cut up the remaining sixths into wedges and regroup them to form a mixed number: 3 whole circles with $\frac{3}{6}$ left over.
- The second student writes the sum and reduces it to lowest terms.
- Pairs repeat the activity, using fraction circles to model the other four problems.

Customizing the Activity for Individual Needs

ESL To reinforce the term *proper fraction*, relate it to everyday phrases using *proper*. "Eating a balanced meal is the *proper* way to eat." "A *proper* fraction has a larger bottom number than top number."

Learning Styles Students can:

 draw a diagram to illustrate like mixed numbers.

Mixed Numbers
$\underbrace{1\frac{2}{3} + 4\frac{1}{3}}_{\text{with like fractions}}$

 demonstrate for the class how to add mixed numbers using fraction circles.

 say aloud the steps needed to add like mixed numbers as a partner writes them.

Enrichment Activity

Have students make charts of weekly practice durations in quarter-hours for different teams or bands. *Example*: Jammers, $3\frac{1}{4}$ hours; Shouters, $2\frac{1}{4}$ hours; Rockers, $4\frac{3}{4}$ hours.

Students can create addition and subtraction word problems. Encourage students to make up varied durations using common denominators. Post the charts and word problems for others to solve.

Alternative Assessment

Students can model addition of like mixed numbers using fraction circles.

Example: Use fraction circles to explain why $1\frac{2}{3} + 1\frac{1}{3} = 3$.

Answer: Divide circles into thirds, regroup $\frac{2}{3} + \frac{1}{3}$ as 1 whole, and then add $1 + 1 + 1 = 3$.

9.3 Subtracting Like Mixed Numbers
Student Edition pages 182–183

Prerequisite Skills
- Changing whole numbers to fractions
- Reducing fractions to lowest terms

Lesson Objectives
- Subtract like mixed numbers.
- Life Skill: Compare distances in miles.

Cooperative Group Activity

Fraction Circles Subtraction

Materials: circles divided into wholes, halves, thirds, fourths, fifths, sixths, eighths, tenths, and twelfths or Visual 7 *Fraction Circles*; markers or colored pencils; scissors

Procedure: Have students work in pairs. Give each student six copies of Visual 7. Write the following problems on the chalkboard:

$2\frac{1}{5} - 1\frac{3}{5}$ $\quad$ $3\frac{1}{6} - 2\frac{5}{6}$

$5\frac{2}{3} - 1\frac{1}{3}$ $\quad$ $4\frac{1}{4} - 2\frac{3}{4}$

Model a problem with the class: $3\frac{3}{4} - 1\frac{1}{4}$.

- One student cuts out 3 whole circles and shades fourths to model $3\frac{3}{4}$, and then he/she places them on a sheet of paper and writes the problem under the models.
- The second student crosses out 1 whole and $\frac{1}{4}$, and then he/she arranges the remaining wholes and fourths to form a mixed number: $2\frac{2}{4}$.
- The first student reduces this difference to lowest terms.
- The class discusses the regrouping that would need to be done if the problem were $3\frac{1}{4} - 1\frac{3}{4}$.
- Pairs repeat the activity for the other four problems using fraction circles to model.

Customizing the Activity for Individual Needs

ESL To help students understand the word *regroup*, explain that the prefix *re-* means *again*. *Regroup* means to group again or in another way. Show how $2\frac{1}{3}$ can be regrouped as $1\frac{4}{3}$.

Learning Styles Students can:

 draw fraction-bar models to subtract like mixed numbers.

 cut out fraction circles and sections and remove wholes and parts to model subtraction of like mixed numbers.

 say aloud the steps needed to subtract like mixed numbers as a partner writes them.

Reteaching Activity

Provide students with fraction models for wholes, halves, thirds, fourths, fifths, sixths, eighths, tenths, and twelfths or three copies of Visual 8 *Fraction Bars*. Have students use the fraction bars to subtract each of the following. Then have students explain the regrouping needed to subtract, using pencil and paper.

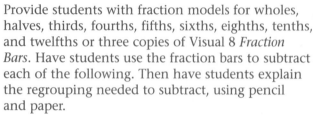

Alternative Assessment

Students can model subtraction of like mixed numbers using fraction circles.

Example: Use fraction circles to show why $3\frac{1}{6} - 1\frac{5}{6} = 1\frac{1}{3}$. How do you regroup?

Answer: Shade 3 whole circles and $\frac{1}{6}$ of a circle. Divide 1 whole circle into sixths to regroup $3\frac{1}{6}$ as $2\frac{7}{6}$. Cross out $1\frac{5}{6}$ circles, leaving $1\frac{2}{6}$ circles. $1\frac{2}{6}$ can be renamed as $1\frac{1}{3}$.

9.4 Subtracting from a Whole Number
Student Edition pages 184–185

Prerequisite Skill
- Changing whole numbers to fractions

Lesson Objectives
- Subtract mixed numbers from whole numbers.
- Subtract fractions from whole numbers.
- Life Skill: Compare weights of foods.

Cooperative Group Activity

Subtract the Fraction

Materials: slips of paper, pencils, paper. *Options*: Visual 1 *Number Lines: Whole Numbers*, Visual 9 *Fraction Cards*, Visual 10 *Mixed Number Cards*

Procedure: Organize students into pairs. Give each pair 19 slips of paper.

- One partner writes 10 different proper fractions on separate slips of paper. The other partner writes the whole numbers 1–9 on separate slips of paper. Students mix their slips of papers and then turn them upside down in a pile.
- Partners take turns picking a fraction and a whole number. For each pick, one partner finds the difference between the whole number and the fraction.

Variation: To form a mixed number, students can use the fraction and whole number chosen; then they should subtract it from 10.

Customizing the Activity for Individual Needs

ESL Use circle models to reinforce the meaning of the words *whole number*, *mixed number*, and *fraction*. Have students name each type of number as you model the different number types.

Learning Styles Students can:

 use fraction circles to cross out the sections to be subtracted from the whole.

 cut out and arrange fraction circle sections to model subtracting from a whole number.

 say the steps as they subtract fractions from a whole number.

Reteaching Activity

Provide students with circles divided into wholes, halves, thirds, fourths, fifths, sixths, eighths, tenths, and twelfths or copies of Visual 7 *Fraction Circles*. Model how to find $4 - \frac{7}{12}$. Cut out 4 whole circles and 1 circle sectioned into twelfths. Replace 1 whole with the twelfths circle, and then cut out 7 sections from the twelfths circle. Show that 3 wholes and $\frac{5}{12}$ remain. Have students make up similar problems and model with fraction circles.

Alternative Assessment

Students can model subtraction of a fraction from a whole number using fraction circles.

Example: Use fraction circles to show why $2 - \frac{5}{6} = 1\frac{1}{6}$.

Answer: Shade 2 whole circles. Divide 1 whole circle into sixths to regroup 2 as $1\frac{6}{6}$. Then cross out $\frac{5}{6}$ of the circle, leaving $1\frac{1}{6}$.

9.5 Adding Unlike Fractions
Student Edition page 186

Prerequisite Skills
- Finding common denominators
- Reducing fractions to lowest terms
- Changing fractions to mixed numbers

Lesson Objective
- Add unlike fractions.

Words to Know
unlike fractions

Cooperative Group Activity

Fraction Bar Addition

Materials: fraction strips for wholes, halves, thirds, fourths, fifths, sixths, eighths, tenths, and twelfths or Visual 8 *Fraction Bars*; markers or crayons; scissors

Procedure: Have students work in pairs. Give each pair four copies of Visual 8. Have students cut out the fraction bars from the visual. Write these problems on the chalkboard.

$\frac{2}{3} + \frac{1}{6}$ $\qquad$ $\frac{3}{8} + \frac{3}{4}$ $\qquad$ $\frac{1}{2} + \frac{3}{5}$ $\qquad$ $\frac{7}{16} + \frac{1}{8}$

Model a problem with the class: $\frac{2}{3} + \frac{3}{4}$.

- One partner shades the thirds bar to show $\frac{2}{3}$. The other partner shades the fourths bar to show $\frac{3}{4}$. Each student then tries to find a fraction bar that can be used to rename both $\frac{2}{3}$ and $\frac{3}{4}$. This fraction bar is twelfths.

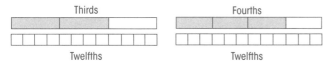

- Each student renames his or her fraction, and together they find the sum.

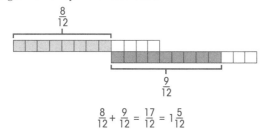

- Pairs repeat the activity using the fraction bars to model the four problems on the chalkboard.

Extension: Students create practice problems of their own in which they add unlike fractions.

Customizing the Activity for Individual Needs

ESL Help students understand the words *unlike fractions* by pointing out that the bottom numbers in two unlike fractions are *unlike* or different.

Learning Styles Students can:

 make lists of multiples for the denominators of the unlike fractions. Then point to the least common multiple in both lists to find the common denominator.

 manipulate fraction bars to find the common denominator.

 say the multiples of each denominator out loud as they generate written lists to find the common denominator.

Enrichment Activity

Have students play the following game in groups of four or five. Students write all of the fractions with denominators of 1, 4, 8, and 16 on slips of paper. Then they put the slips of paper in a bag. One student draws two fractions. The first student in the group to find the sum gets 1 point. Students take turns drawing the fractions. When the game is over, the student with the most points is the winner.

Alternative Assessment

Students can model addition of unlike fractions using fraction circles.

Example: Use fraction circles to show why $\frac{2}{3} + \frac{5}{6} = 1\frac{1}{2}$.

Answer: Shade $\frac{2}{3}$ on a thirds circle and rename it as $\frac{4}{6}$ by dividing it into sixths. Then shade $\frac{5}{6}$ on a sixths circle and add, resulting in $1\frac{1}{2}$.

MATH IN YOUR LIFE
Pay Day
Student Edition page 187

Lesson Objectives
- Apply adding and subtracting fractions to find work hours.
- Life Skill: Calculate pay given time and rate.
- Communication Skill: Write a letter to a temporary employment agency.
- Communication Skill: Explain a time card to an employee or employer.

Activities
Tools of the Trade
Materials: copies of time cards shown below

Procedure: Give each student a copy of the time card below. Have them complete the time card by filling in the number of hours worked each day using fractions for parts of hours. Explain that overtime is more than 40 hours in a week.

Name: Colby Smith			
Day	In	Out	Hours
Mon.	7:00 A.M.	2:00 P.M.	7
Tues	11:00 A.M.	5:15 P.M.	$6\frac{1}{4}$
Wed.	8:30 A.M.	6:15 P.M.	$9\frac{3}{4}$
Thurs.	8:15 A.M.	4:00 P.M.	$7\frac{3}{4}$
Fri.	2:00 P.M.	9:30 P.M.	$7\frac{1}{2}$

Then have students determine:
- the total number of hours that Colby worked this week.
- whether or not Colby qualifies for overtime this week and then explain why.

Write a Business Letter
Materials: paper and pens or computer word processor

Procedure: Have students write letters to various temporary employment agencies. In their letters, have them ask:
- how workers report hours.
- how fractional parts of hours are reported.
- how workers are paid for fractional parts of an hour.

When responses come in, have students compare the different reporting systems. Have students decide which temporary agency they would like to work for and explain why.

Act It Out
Materials: completed time card, such as the one from "Tools of the Trade" above

Procedure: Have students role play an employer and an employee who thinks his or her paycheck is incorrect based on the hours that he or she worked. Have the employee gather evidence to support his or her position. For example, the employee would need:
- a copy of the paycheck and pay stub.
- a copy of his or her time card showing the hours worked.

Have other students suggest different ways the employer might handle the situation.

Practice
Have students complete Math in Your Life 115 *Pay Day* from the Classroom Resource Binder.

Internet Connection
Students can use the following temporary agencies' Web sites to find out about their policies:
Kelly Services
    http://www.kellyservices.com
Olsten Staffing Services
    http://www.olsten.com

9.6 Subtracting Unlike Fractions
Student Edition page 188

Prerequisite Skills
- Finding common denominators
- Subtracting like fractions
- Reducing fractions to lowest terms

Lesson Objective
- Subtract unlike fractions.

Cooperative Group Activity
Fraction Bar Subtraction
Materials: fraction strips for wholes, halves, thirds, fourths, fifths, sixths, eighths, tenths, and twelfths or Visual 8 *Fraction Bars*; markers or crayons; scissors

Procedure: Have students work in pairs. Give each pair four copies of Visual 8. Have students cut out the fraction bars from the visual. Write these problems on the board, and then model the first problem.

$\frac{9}{10} - \frac{2}{5}$ $\frac{5}{8} - \frac{1}{4}$ $\frac{5}{6} - \frac{1}{3}$ $\frac{7}{8} - \frac{3}{4}$

Model a problem with the class: $\frac{7}{10} - \frac{1}{5}$.

- Pairs use the tenths bar and the fifths bar. They shade the bars to show $\frac{7}{10}$ and $\frac{1}{5}$. They then line up the shaded part of the fifths bar above the shaded part of the tenths bar. They notice that $\frac{1}{5} = \frac{2}{10}$. They cross out $\frac{2}{10}$ and reduce the result to lowest terms.

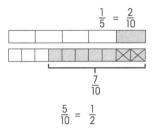

- Pairs repeat the activity using the fraction bars to model the four problems on the chalkboard.

Extension: Students can create practice problems of their own in which they subtract unlike fractions.

Customizing the Activity for Individual Needs

ESL Remind students that *difference* means the answer to a subtraction problem. Write subtraction examples with whole numbers on the chalkboard, and have students label the *difference*. Have them point to the *difference* as they solve the problems.

Learning Styles Students can:

 make lists of multiples for the denominators of the unlike fractions. Then they should point to the least common multiple in both lists to find the common denominator.

 manipulate fraction bars to find the common denominator.

 say the multiples of each denominator out loud as they generate written lists to find the common denominator.

Enrichment Activity

Challenge student pairs to arrange each set of numbers in the boxes below to write two correct subtraction problems. Have them give the problems to each other to solve.

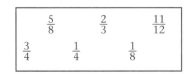

Answers: $\frac{11}{12} - \frac{1}{4} = \frac{2}{3}$; $\frac{3}{4} - \frac{1}{8} = \frac{5}{8}$

| | $\frac{5}{6}$ | $\frac{1}{9}$ | $\frac{2}{3}$ |
| $\frac{1}{3}$ | $\frac{4}{9}$ | | $\frac{1}{6}$ |

Answers: $\frac{4}{9} - \frac{1}{3} = \frac{1}{9}$; $\frac{5}{6} - \frac{2}{3} = \frac{1}{6}$

Alternative Assessment

Students can model subtraction of unlike fractions using fraction circles.

Example: Use fraction circles to show why $\frac{7}{8} - \frac{1}{4} = \frac{5}{8}$.

Answer: Shade $\frac{7}{8}$ on the eighths circle and $\frac{1}{4}$ on the fourths circle. Cut out the shaded $\frac{1}{4}$ and place it on the shaded $\frac{7}{8}$ to show that $\frac{1}{4} = \frac{2}{8}$. Cross out $\frac{2}{8}$, leaving $\frac{5}{8}$.

USING YOUR CALCULATOR
Finding Common Denominators
Student Edition page 189

Lesson Objectives
- Use a calculator to find multiples for a common denominator.
- Add and subtract unlike fractions.

Activity

More or Less Than 1?

Materials: fractions written on slips of paper or Visual 9 *Fraction Cards*, calculators

Procedure: Organize students into pairs. Place 20 different proper fractions into a paper bag.

- Students take turns picking two fractions. On each turn, they estimate whether the sum will be more or less than 1.
- Check each estimate using a calculator to find the least common denominator.

For example, $\frac{2}{3} + \frac{3}{4}$.

Find multiples of 3.

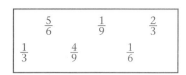

Find multiples of 4.

PRESS [4]+[4]= [8]
PRESS [+][4]= [12]

(Continued on p. 110)

(Continued from p. 109)

$M_3 = \{3, 6, 9, \underline{12}\}$
$M_4 = \{4, 8, \underline{12}\}$
LCD = 12

- Make like fractions, add, and compare to 1.
$\frac{2}{3} + \frac{3}{4} = \frac{8}{12} + \frac{9}{12} = \frac{17}{12}$. An improper fraction is always greater than 1.

Variation: Students estimate whether the difference between the larger and smaller fractions will be more or less than $\frac{1}{2}$.

9.7 Adding Unlike Mixed Numbers
Student Edition pages 190–191

Prerequisite Skills
- Finding common denominators
- Reducing fractions to lowest terms
- Changing fractions to mixed numbers

Lesson Objectives
- Add unlike mixed numbers.
- Life Skill: Add fractional measurements.

Cooperative Group Activity

Adding Ingredients

Materials: cookbooks, cooking magazines, or other magazines and newspapers containing recipes; pencils; paper

Optional: measuring utensils, mixing bowls, rice or sand

Procedure: Organize students into groups of three. Give each group a cookbook or several recipes. Assign each group a different mixing bowl capacity: for example, 2 cups, 3 cups, 4 cups, 6 cups, and so on.

- Each group chooses a recipe with fractional amounts of ingredients. Group members work together to find the total capacity of the ingredients in their recipe.
- The students decide whether their assigned mixing bowl is large enough to hold all of the ingredients. If not, they tell what size mixing bowl they would need.

Variations: Provide mixing bowls and measuring utensils with rice or sand so that students can test their assumptions.

Customizing the Activity for Individual Needs

ESL To review *unlike mixed numbers*, have students orally identify parts of a labeled diagram.

Mixed Numbers

$2\frac{3}{4} + 5\frac{1}{8}$
  └─── unlike fractions ───┘

Learning Styles Students can:

 highlight the unlike fractions and rename to add.

 use manipulatives to add unlike mixed numbers.

 first say the steps aloud as they rename the unlike fractions and then add.

Enrichment Activity

Have students replace each blank below with a number that will make the addition sentence true. Then have them make up their own exercises for classmates to solve.

$1\frac{1}{4} + \underline{} = 2\frac{1}{2}$

$7\frac{1}{6} + \underline{} = 10\frac{2}{3}$

$4\frac{5}{8} + \underline{} = 7\frac{1}{8}$

$2\frac{3}{5} + \underline{} = 4\frac{9}{10}$

Alternative Assessment

Students can model addition of unlike mixed numbers using fraction circles.

Example: Use fraction circles to show why $1\frac{3}{4} + 1\frac{7}{8} = 3\frac{5}{8}$. Why is the sum greater than 3?

Answer: Shade $1\frac{3}{4}$ and $1\frac{7}{8}$ circles. Rename $\frac{3}{4}$ as $\frac{6}{8}$ by dividing the fourths into eighths. Then combine all the shaded parts. The sum is greater than 3 because the sum of the fraction parts of the mixed numbers is greater than 1.

9.8 Subtracting Unlike Mixed Numbers
Student Edition pages 192–193

Prerequisite Skills
- Finding common denominators
- Changing whole numbers to fractions
- Reducing fractions to lowest terms

Lesson Objectives
- Subtract unlike mixed numbers.
- Life Skill: Subtract to compare fractional measurements and prices.

Cooperative Group Activity

Buying Fabric
Materials: paper, pencils

Procedure: Organize students into pairs. Have pairs role play the owner of a fabric store and a customer. Each pair begins with a roll of fabric 20 yards in length. The standard width for the pieces of fabric is 1 yard.

- The customer tells the owner how much fabric he or she needs, using a mixed number. For example: "I need $3\frac{7}{8}$ yards of fabric in length." The owner must keep a record of how much fabric is left. He or she subtracts the amount of fabric and finds the difference.
- After each *sale*, the owner and customer switch roles. Pairs continue to keep track of the remaining fabric until the roll is used up.

Customizing the Activity for Individual Needs
ESL Help students to distinguish between *rename* and *regroup* for a mixed number. Point out that when you rename, you find another name for the fraction. When you regroup, you group part of the whole number as a fraction.

$$\underbrace{2\frac{1}{2} = 2\frac{2}{4}}_{\text{rename}} \qquad \underbrace{2\frac{2}{4} + 1\frac{6}{4}}_{\text{regroup}}$$

Learning Styles Students can:

 highlight in one color the whole number part of a mixed number and in another color the fraction part.

 use manipulatives to subtract unlike mixed numbers.

 say the steps as they regroup the whole number part of a mixed number to subtract.

Enrichment Activity
Have students solve the problem below. Then have students make up similar problems for classmates to solve.

> A carpenter started with a 12-foot-long wooden beam. She cut pieces $4\frac{3}{8}$ feet long and $5\frac{1}{6}$ feet long. How much of the original beam is left?

Answer: $12 - 4\frac{3}{8} = 7\frac{5}{8}$; $7\frac{5}{8} - 5\frac{1}{6} = 2\frac{11}{24}$

Alternative Assessment
Students can model subtraction of unlike mixed numbers using fraction circles.

Example: Use fraction circles to show why $2\frac{1}{3} - 1\frac{5}{6} = \frac{1}{2}$. What fraction do you rename? What number do you regroup?

Answer: Rename $\frac{1}{3}$ as $\frac{2}{6}$; shade sixths circles and regroup $2\frac{2}{6}$ as $1\frac{8}{6}$. Cross out $1\frac{5}{6}$ circles, leaving $\frac{3}{6}$ or $\frac{1}{2}$ circle.

9.9 Problem Solving: Multi-Part Problems
Student Edition pages 194–195

Prerequisite Skills
- Recognizing clue words
- Solving two-part problems
- Adding, subtracting, and multiplying

Lesson Objectives
- Solve multi-part word problems involving fractions and mixed numbers
- Life Skill: Calculate lengths of fabric.

Cooperative Group Activity

Step-by-Step
Materials: outline of four problem-solving steps (Read, Plan, Do, Check) or Visual 22 *Problem Solving Steps*; pencils; paper

Procedure: Organize students into an even number of small groups. Give each group several copies of Visual 22.

- Groups use the visual to work out each part of any multi-part problems from the Student Edition, Workbook, or Classroom Resource Binder, such as Practice 113 *Problem Solving: Multi-Part Problems*.
- Group members work together to make up their own multi-part problems. They write the problems and their solutions on separate sheets of paper. Then they trade problems with another group.
- Students use each other's solution sheets to check their answers.

Customizing the Activity for Individual Needs

ESL To help students recognize implied questions in a multi-part problem, write out the two questions to be answered in the PLAN step on Visual 22 *Problem Solving Steps*.

Learning Styles Students can:

 highlight clue words for different operations in different colors to determine the number of steps and operations to use.

 write facts given in a problem on separate index cards. They then can group the cards needed to solve each part of the problem.

 read the steps on Visual 22 aloud as they fill them in.

Enrichment Activity

Have students use practice problems they've solved to make up *Matching Missing Number* problems. They can write one problem on each side of an index card. The same missing number must work in both problems, using different operations.

Example:

$$\begin{array}{r} 6\frac{3}{4} \\ -\blacksquare \\ \hline 5\frac{3}{4} \end{array} \qquad \frac{3}{4} \times \blacksquare = \frac{3}{4}$$

Answer: ■ = 1

Encourage students to solve each other's problems.

Alternative Assessment

Students can explain a multi-step problem.

Example: What steps are needed to solve this problem? What operations should you use?

Jake uses $1\frac{1}{2}$ yards of red and $2\frac{1}{4}$ yards of white fabric to make a tablecloth. The fabric costs $4 a yard. How much will the tablecloth cost?

Possible Answer: Find the total amount of fabric; find the cost. Add to find the total amount of fabric. Multiply to find the total cost.

Closing the Chapter
Student Edition pages 196–197

Chapter Vocabulary

Review with students the Words to Know on page 177 of the Student Edition. Then have students quiz each other in pairs.

Have students copy and complete the Vocabulary Review questions on page 196 of the Student Edition.

For more vocabulary practice, have them complete Vocabulary 101 *Adding and Subtracting Fractions* from the Classroom Resource Binder.

Test Tips

Have students read each Test Tip and then pick out the problems in which they can apply that tip. Encourage students to explain how the tip will help them with a particular type of problem.

Learning Objectives

Have students review CM6 *Goals and Self-Check* from the Classroom Resource Binder. They can check off the goal they have reached. Note that each section of the quiz corresponds to a Learning Objective.

Group Activity

Summary: Students monitor and record the prices of five stocks over the course of five days to find their profit and loss each day.

Materials: Business sections of newspapers, pencils, paper

Procedure: Allow time each day for students to review the business section or to use the Internet to check stock prices. Or you may assign this as homework.

Assessment: Use the Group Activity Rubric on page *xii* of this guide. Fill in the rubric with the additional information below. For this activity, students should have:

- selected five stocks and found the cost of 100 shares.
- recorded daily closing prices and profit or loss.

RELATED MATERIALS See the unit overview page for other Globe Fearon books that can be used to enrich and extend the material in this unit.

Assessing the Chapter

Traditional Assessment

Chapter Quiz
The Chapter Quiz on pages 196–197 of the Student Edition can be used as an open- or closed-book test, or as homework. The quiz can be used to identify concepts in the chapter that students need to review and practice.

Chapter Tests
Use Chapter Test A Exercise 117 and Chapter Test B Exercise 118 *Adding and Subtracting Fractions* from the Classroom Resource Binder to further assess mastery of chapter concepts.

Additional Resources
Use the Resource Planner on pages 101–102 to assign additional exercises from the Classroom Resource Binder and Workbook.

Alternative Assessment

Interview
Write $6\frac{1}{6} - 2\frac{3}{4}$ on the board. Ask:
- How do you know if you will need to regroup in this problem?
- Why do you need a common denominator for this problem?
- Why can't you use 6 as the common denominator?
- Will the difference be greater or less than 4?

Presentation
Write the numbers $4\frac{2}{3}$, $1\frac{5}{6}$, and $3\frac{5}{12}$ on the chalkboard. Ask students to demonstrate how they would do each of the following:
- Find a common denominator to add any two of the given numbers.
- Model addition with fraction circles (Visual 7).
- Model subtraction with fraction bars (Visual 8).
- Write a multi-part problem using at least two of the given numbers.
- Solve the problem.

Unit Assessment

This is the last chapter in Unit 2, Fractions. To assess cumulative knowledge and provide standardized-test practice, administer the practice test on page 198 of the Student Edition and the Unit 2 Cumulative Test, page T9, in the Classroom Resource Binder. These tests are in multiple-choice format. A scantron sheet is provided on page T2 of the Classroom Resource Binder.

Unit Overview

Unit 3 ▶ Other Types of Numbers

CHAPTER 10 Decimals	**PORTFOLIO PROJECT** Decimal Search	**ON-THE-JOB MATH** Bank Teller	**USING YOUR CALCULATOR** Number Patterns	**PROBLEM SOLVING** Multi-Part Decimal Problems
CHAPTER 11 Percents	**PORTFOLIO PROJECT** Nutrition Facts	**MATH IN YOUR LIFE** Tipping	**USING YOUR CALCULATOR** The Percent Key	**PROBLEM SOLVING** Finding the Part, Percent, or Whole
CHAPTER 12 Ratios and Proportions	**PORTFOLIO PROJECT** Scale Drawing	**ON-THE-JOB MATH** Courier	**USING YOUR CALCULATOR** Solving Proportions	**PROBLEM SOLVING** Using Proportions

RELATED MATERIALS

These are some of the Globe Fearon books that can be used to enrich and extend the material in this unit.

Practice & Remediation ▶

Access To Math
Provide your students with the reinforcement practice they need to develop basic math proficiency.

Test Preparation ▶

Math for Proficiency Level B
Give your students the skills needed to succeed on proficiency exams through this complete test taking text.

Real-World Math ▶

Consumer Math
Bring core math skills to life in your class with this rich consistent program.

Planning the Chapter

Chapter 10 • Decimals

Chapter at a Glance

SE page	Lesson	
200		*Chapter Opener and Project*
202	10.1	What Is a Decimal?
203	10.2	Reading and Writing Decimals
205	10.3	Comparing Decimals
207	10.4	Ordering Decimals
208	10.5	Adding Decimals
210	10.6	Subtracting Decimals
212	10.7	Multiplying Decimals
214	10.8	Multiplying Decimals by 10, 100, 1,000
215		*On-The-Job Math: Bank Teller*
216	10.9	Dividing Decimals by Whole Numbers
218	10.10	Dividing Decimals by Decimals
220	10.11	Dividing Decimals by 10, 100, 1,000
221		*Using Your Calculator: Number Patterns*
222	10.12	Problem Solving: Multi-Part Decimal Problems
224	10.13	Renaming Decimals as Fractions
225	10.14	Renaming Fractions as Decimals
226	10.15	Rounding Decimals
228		*Chapter Review and Group Activity*

Learning Objectives

- Read and write decimals.
- Compare and order decimals.
- Add, subtract, multiply, and divide decimals.
- Change a decimal to a fraction.
- Change a fraction to a decimal.
- Round decimals.
- Solve problems using decimals.
- Apply knowledge of decimals to banking.

Life Skills

- Locate library books.
- Find total costs.

Communication Skills

- Read decimals in newspapers and magazines.
- Explain how to solve a problem.
- Write a letter.

Workplace Skills

- Check deposit slips.
- Help customers with their banking needs.

PREREQUISITE SKILLS

To assess mastery of prerequisite skills, use a selection of exercises from any of the program resources referenced below. The same resources can be used to provide remediation if necessary.

Program Resources for Review

Skills	Student Edition	Workbook	Classroom Resource Binder
Using whole number place value	Lesson 1.3	Exercises 3, 4	Review 2
Reading and writing whole numbers	Lesson 1.5	Exercise 5	Practice 5
Comparing and ordering whole numbers	Lessons 1.6, 1.7	Exercises 6, 7	Practice 6
Adding whole numbers with regrouping	Lessons 2.6, 2.7	Exercises 15, 16	Practice 17
Subtracting whole numbers with regrouping	Lessons 3.5, 3.6	Exercises 22, 23	Practice 27, 28
Multiplying whole numbers with regrouping	Lessons 4.4, 4.5, 4.7	Exercises 29, 30, 32	Practice 39, 40
Dividing whole numbers	Lessons 5.3–5.8	Exercises 37–42	Practice 56

Skills	Program Resources for Review		
	Student Edition	Workbook	Classroom Resource Binder
Solving two-part problems	Lesson 4.9	Exercise 34	Practice 43
Simplifying fractions	Lesson 7.3	Exercise 56	Review 78
Rounding whole numbers	Lesson 1.9	Exercise 9	Practice 4

Diagnostic and Placement Guide

Chapter 10 Decimals

The Percent Accuracy scores are based on the number of problems in a lesson that have been answered correctly.

Resource Planner

After diagnosing your students' needs, use the Program Resources for reinforcement, reteaching, or enrichment. Activities for customizing lessons can be found in this guide on the pages shown.

KEY
Reteaching = ⤺   Reinforcement = ⬇   Enrichment = ⤻

Lessons	Percent Accuracy				Program Resources			
	50%	65%	80%	100%	Student Edition Extra Practice	Workbook Exercises	Teacher's Planning Guide	Classroom Resource Binder
10.1 What Is a Decimal?	8	10	12	15		84	⤺ p. 118	Visual 13 *Decimal Models*
10.2 Reading and Writing Decimals	7	9	11	14	p. 429	85	⬇ p. 119	⬇ Review 120 *Reading and Writing Decimals* Visual 13 *Place-Value Chart: Decimals*
10.3 Comparing Decimals	7	9	11	12	p. 429	86	⬇ p. 119	⬇ Practice 126 *Comparing and Ordering Decimals*
10.4 Ordering Decimals	8	10	13	16	p. 429	87	⬇ p. 120	⬇ Practice 126 *Comparing and Ordering Decimals*
10.5 Adding Decimals	10	13	16	20	p. 430	88	⤺ p. 121	⬇ Review 121 *Adding Decimals* ⬇ Practice 127 *Adding Decimals* Visual 14 *Place-Value Study Tool: Decimals* Visual 13 *Decimal Models*
10.6 Subtracting Decimals	10	13	16	20	p. 430	89	⤺ p. 121	⬇ Review 122 *Subtracting Decimals* ⬇ Practice 128 *Subtracting Decimals* ⬇ Practice 132 *Adding and Subtracting Decimals*
10.7 Multiplying Decimals	11	14	17	21	p. 430	90	⤺ p. 122	⬇ Review 123 *Multiplying Decimals* ⬇ Practice 129 *Multiplying Decimals*

Lessons		Percent Accuracy				Program Resources			
		50%	65%	80%	100%	Student Edition Extra Practice	Workbook Exercises	Teacher's Planning Guide	Classroom Resource Binder
10.8	Multiplying Decimals by 10, 100, 1,000	6	9	10	12		91	p. 123	
	On-The-Job Math: Bank Teller	Can be used for portfolio assessment.						p. 124	On-The-Job Math 133 Bank Teller
10.9	Dividing Decimals by Whole Numbers	10	13	16	20	p. 431	92	p. 124	Review 124 Dividing Decimals by Whole Numbers
10.10	Dividing Decimals by Decimals	14	18	22	28	p. 431	93	p. 125	Review 125 Dividing Decimals by Decimals
10.11	Dividing Decimals by 10, 100, 1,000	6	9	10	12		94	p. 126	Practice 130 Dividing Decimals
10.12	Problem Solving: Multi-Part Decimal Problems	1	2	2	3	p. 431	95	p. 127	Practice 131 Problem Solving: Multi-Part Decimal Problems Visual 22 Problem Solving Steps
10.13	Renaming Decimals as Fractions	10	13	16	20	p. 432	96	p. 127	
10.14	Renaming Fractions as Decimals	13	16	20	25	p. 432	97	p. 128	Challenge 134 Pack It Up
10.15	Rounding Decimals	16	21	26	32		98	p. 128	
Chapter 10 Review									
	Vocabulary Review	4	5.5	6.5	8	p. 432		p. 129	Vocabulary 119
		(writing is worth 4 points)							
	Chapter Quiz	18	24	29	36			p. 130	Chapter Tests A, B 135, 136

Customizing the Chapter

Opening the Chapter
Student Edition pages 200–201

Photo Activity
Procedure: Discuss athletic events with students, and talk about how decimals are used in sports. As a motivational activity, you may have students discuss sports competitions they have seen in which there were only seconds left for a team to win. Or discuss favorite players' statistics to introduce decimals with students. Provide relevant visuals, such as baseball cards with statistics, the sports section of the newspaper, or an article about a track and field competition.

Words to Know
Review with students the Words to Know on page 201 of the Student Edition. You may wish to spend more time on words that are unfamiliar to your class.

The following words and definitions are covered in this chapter.

decimal a number that names part of a whole

decimal point the dot in a decimal; a decimal has digits to the right of its decimal point

decimal places the places to the right of a decimal point

mixed decimal a number with a whole number and a decimal

Decimal Search Project

Summary: Students collect examples of decimals in newspapers, magazines, or textbooks and then make up problems.

Materials: newspapers, magazines, sports cards, food packages, sales receipts

Procedure: Provide the materials for students to complete the projects in class. This project may also be used as a daily homework assignment as students complete the chapter.

Assessment: Use the Individual Activity Rubric on page *xi* of this guide. Fill in the rubric with the additional information below. For this project, students should have:

• explained what the decimals describe.
• written two problems using their decimals.

Learning Objectives

Review with students the Learning Objectives on page 201 of the Student Edition before starting the chapter. Students can use the list of objectives as a learning guide. Suggest that they write the objectives in a journal or use CM6 *Goals and Self-Check*.

After each lesson, have students write an example of the skill they learned under the appropriate objective. Suggest that students use these worksheets as a learning guide to help them study for the chapter test.

10.1 What Is a Decimal?
Student Edition page 202

Prerequisite Skill
• Understanding parts of a whole

Lesson Objective
• Identify decimals and their parts.

Words to Know
decimal, decimal point, decimal places, mixed decimal

Cooperative Group Activity

Deck of Decimals

Materials: decimal grids (graph paper cut into 10 x 10 cards) or copies of Visual 11 *Tenths Models* and Visual 12 *Hundredths Models* reduced in size.

Procedure: Organize the class into pairs. Distribute six grids or three copies of each model and six cards to each pair of students.

• One student shades a grid to model a tenths or hundredths decimal. The other student writes the decimal on a blank index card. Then students switch roles until all grids and cards are filled in.

• Pairs collect cards, shuffle them, and turn them facedown. Students take turns flipping over two cards at a time. If a decimal matches its model, the player keeps the cards.

• Play continues until all cards are matched. The player with the most cards wins. Have pairs trade decks and play again.

Customizing the Activity for Individual Needs

ESL Help students understand the word *decimal*. Write a decimal on the chalkboard. Say each part out loud as you label it. Have students count the number of digits after the decimal aloud in their native languages. Then have students count aloud in English.

Learning Styles Students can:

 strike through each decimal place as they count the number of digits.

 touch the digits as they count them.

 count out loud the number of places after the decimal point.

Reteaching Activity

Write the numbers .1 to .9 each on a separate index card. Have a student pick a card. Then have the student shade the appropriate number of boxes on Visual 11 *Tenths Model* to represent the number. Repeat the activity with hundredths for any ten decimal numbers, using Visual 12 *Hundredths Model*.

Alternative Assessment

Students can identify the number of places after a decimal point by highlighting the digits.

Example: Tell the number of decimal places in 14.567.

Answer: There are three decimal places.

10.2 Reading and Writing Decimals *Student Edition pages 203–204*

Prerequisite Skills
- Understanding whole-number place value
- Reading and writing whole numbers

Lesson Objectives
- Read decimals.
- Write decimals.
- Communication Skill: Read decimals in newspapers.

Cooperative Group Activity

Two Ways
Materials: index cards, pencils, paper

Procedure: Organize students into pairs. Give each student four index cards. Have students write a decimal in numerical form and in word form on one side of each card. Collect the cards and shuffle them. Place them facedown in a pile. Have each pair pick four index cards from the pile.
- One partner draws a card and reads the number on it aloud without his or her partner seeing it. The partner writes the number in numerical form on a sheet of paper.
- Partners compare what is written with what is on the index card.
- Partners switch roles and repeat.
- Students then repeat the activity, writing the number in word form.

Customizing the Activity for Individual Needs
ESL To help students understand the meaning of the words *thousand* and *thousandth*, write numbers that illustrate each word. Then have students read each number.

Example: 5,000   5 thousand
             .005   5 thousandth

Repeat for *hundred* and *hundredth*, and for *ten* and *tenth*. Students may work in pairs to review these number words.

Learning Styles Students can:

write the number on a place-value chart and then highlight each column after the decimal with a different color.

write the number on a place-value chart, then put one finger on the last digit, and finally drag the finger up to determine the place value of the digit.

read the decimal number aloud before writing the number.

Reinforcement Activity

Provide students with four number cubes, each numbered 1–6. Students roll the cubes and write a decimal number using the numbers rolled. They then write the number in word form. Challenge them to write different numbers by moving the decimal point.

Alternative Assessment

Students can match a decimal written in word form with its numerical form.

Example: Which decimal form represents sixty and two hundredths?

   .62   60.2   60.02   60.002

Answer: 60.02

10.3 Comparing Decimals *Student Edition pages 205–206*

Prerequisite Skills
- Comparing whole numbers
- Reading and writing decimals

Lesson Objectives
- Compare two decimals.
- Life Skill: Locate library books.

Cooperative Group Activity

Roll a Number
Materials: number cubes, self-stick notes

Procedure: Organize students into groups of three. Give each group four number cubes.
- One member of the group rolls the number cubes. He or she uses those numbers to write a decimal number on a self-stick note.

- A second member of the group also rolls the number cubes. He or she uses those numbers to write a decimal number on a self-stick note.
- The third group member writes a symbol (<, >, or =) on a self-stick note and then arranges the self-stick notes to compare the numbers.
- Group members rotate roles and repeat the activity two more times.

Customizing the Activity for Individual Needs

ESL Be sure that students understand that the symbols >, <, and = stand for the words *greater than*, *less than*, and *equal to*. Have students write each symbol on a separate index card. Then write the words for each symbol next to it, both in English and in students' native languages. Encourage students to use the cards as self-help cards when comparing decimals.

Learning Styles Students can:

 align the digits of the numbers by their place value to see which is greater.

 use Visual 11 *Tenths Model* and/or Visual 12 *Hundredths Model* to shade and cut out decimal parts; then compare the parts to see which is greater.

 read aloud the value of each digit of the two decimal numbers to determine which number is greater.

Reinforcement Activity

Provide students with a copy of Visual 14 *Place Value Study Tool: Decimals* or a decimals place-value chart. Give them several different decimal numbers. Have them write the decimals in the chart. Then have them compare the decimals by comparing the digits in each place, working from left to right.

Alternative Assessment

Students can identify which decimal is greater by using words rather than symbols.

Example: Compare .56 and .506.

Answer: .56 is greater than .506 because .560 > .506.

10.4 Ordering Decimals
Student Edition page 207

Prerequisite Skills
- Ordering whole numbers
- Comparing whole numbers

Lesson Objective
- Order three decimal numbers from largest to smallest.

Cooperative Group Activity

Find the Largest

Materials: large self-stick notes, markers, poster board

Procedure: Make a four-row decimal place-value chart on a large sheet of poster board. Mount it on the wall of the classroom. Use a marker to make decimal points in each of the squares of the middle column. Have students write the digits 0–9 on three sets of separate self-stick notes. Also have students make extra zero notes. Collect the notes. Organize the class into three groups. Place the numbered notes on the grid to make three different decimal numbers.

- One group "sticks" zeros where needed so that the numbers all have the same number of decimal places.
- Another group decides which number is the largest of the three numbers and then explains why.
- The third group decides which of the remaining numbers is the larger and then explains why.
- The notes are collected. Repeat the activity for three new numbers.

Variation: Stick one number to the chart and then challenge groups to stick a smaller number above yours and a larger one below.

Customizing the Activity for Individual Needs

ESL To help students differentiate between *comparing* and *ordering* numbers, point out that you *compare two* numbers but you *order three or more* numbers. Provide an example by arranging decimal models in order from greatest to least.

Learning Styles Students can:

 circle the digits that indicate which number is largest.

 remove self-stick notes in each place value where the digits are the same before deciding which number is largest.

 explain to the class why a chosen number is the largest.

Reteaching Activity

Provide students with blank number lines or Visual 1 *Number Lines: Whole Numbers*. Have them number one of the blank number lines from 0 to 1 in 0.1 increments. Give students three decimal numbers in tenths that are less than 1. Have students place the decimals on the number line to determine their order.

Alternative Assessment

Students can order decimal numbers by creating and ordering the shaded models of the numbers.

Example: Order .03, .25, and .14 from largest to smallest.

Answer: .25, .14, .03

10.5 Adding Decimals
Student Edition pages 208–209

Prerequisite Skills
- Adding whole numbers with regrouping
- Using decimal place value

Lesson Objectives
- Add decimals.
- Life Skill: Add prices to find total cost.
- Life Skill: Find costs using a cash register receipt.

Cooperative Group Activity

The Price Is Right
Materials: play money or Visual 15 *Money Models*; sale items such as video games, CDs, batteries, small electronics items with price tags; paper; pencils

Procedure: Organize the class into small groups. Prepare three envelopes of varying amounts of play money for each group. Assign prices, and attach price tags to each sale item. Give each group some envelopes of play money and several sale items.

- One student is the cashier and does not get an envelope.
- Students individually count the money in the envelopes, then work together to total the money and check the amount.
- The group decides which items to buy, gathers the money, and then gives it to the cashier.
- The cashier checks to see if the sum received covers the cost of the purchase and determines if change is needed.
- Students switch roles using new envelopes and items.

Customizing the Activity for Individual Needs

ESL Pair students to review the words and values of penny, nickel, dime, and quarter. Have them take turns, one identifying a coin while the other says its value in cents and as part of a dollar. For example: "This is a dime. It is worth ten cents or one-tenth of a dollar."

Learning Styles Students can:

 record on paper the prices of items and amounts of money or decimal numbers to be totaled.

 use play money to model problems and find sums.

 count aloud as amounts are added.

Reteaching Activity

Provide students with hundredths grids or Visual 12 *Hundredths Models* to help them develop a concrete understanding of adding decimals.

Alternative Assessment

Students can model addition by shading hundredths grids or Visual 12 *Hundredths Model*.

Example: Use hundredths models to show why .72 + .08 = .80.

Answer: Students shade hundredths grids as shown.

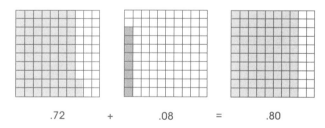

.72 + .08 = .80

10.6 Subtracting Decimals
Student Edition pages 210–211

Prerequisite Skills
- Using whole-number place value
- Subtracting whole numbers with regrouping

Lesson Objectives
- Subtract decimals.
- Compare decimals.
- Life Skill: Keep a check register.

Cooperative Group Activity

Check Registers

Materials: blank check registers in a checkbook, sale items with price tags, paper, pencils

Procedure: Write a different balance at the top of each check register. Organize students into groups of four or five. Give each student a check register. Attach price tags with prices to different items in the classroom that will be used as sale items. Set up one shopping station per group.

- Each student chooses an item to buy. The student writes the name of the item and the cost in the check register; then he or she subtracts to find the new balance.
- Students repeat buying items until their balances are too small to buy any more items.
- Groups check each other's balances. Then they total how much money they have left as a group.
- Groups decide what to buy with their totals so that their balances are as close to $0 as possible.
- Groups switch shopping stations and repeat the activity with new balances on their check registers.

Customizing the Activity for Individual Needs

ESL Explain that the *balance* is the amount of money in the checking account at any one time. Point out the balance on a check register, and explain how the balance was found.

Learning Styles Students can:

 highlight purchase costs in their check registers as a reminder to subtract.

 use play money to model purchases and decreases in their balances.

 tell group members their original balances, the cost of each item purchased, and the remaining balances after each purchase.

Reteaching Activity

Provide students with hundredths grids or Visual 12 *Hundredths Model* to help them develop a concrete understanding of subtracting decimals.

Alternative Assessment

Students can model subtraction with decimals by shading hundredths grids.

Example: Use hundredths models to show why .72 − .08 = .64.

Answer: Students shade hundredths grids as shown.

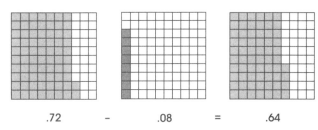

.72 − .08 = .64

10.7 Multiplying Decimals
Student Edition pages 212–213

Prerequisite Skills
- Multiplying whole numbers with regrouping
- Using decimal place value

Lesson Objectives
- Multiply decimals.
- Life Skill: Multiply prices to find total costs of items purchased.
- Communication Skill: Interpret information in an advertisement.

Cooperative Group Activity

Supermarket Dash

Materials: supermarket advertisements, paper, pencils

Procedure: Organize students into groups of three. Create a shopping list from an advertisement. Assign a quantity to each item on the list. For example: Apples—2.75 pounds. Supply enough copies of the list for each group or put the list on the chalkboard. Give each group a copy of the supermarket advertisement. Tell students they are to go shopping for all the items on their lists.

- One student is responsible for locating the items in the advertisement.
- The student reads off the prices to the group members, who record the prices on a chart.
- Group members help one another to figure out the total cost of each item. As a bonus, the group can review adding decimals by calculating the total cost of their shopping spree.
- When a group is done, a student raises his or her hand.
- Checking the answers can be a whole-class activity.

Customizing the Activity for Individual Needs

ESL In order to place decimal points correctly in answers, students must understand the meaning of "count from right to left" and "decimal places." Have students draw arrows from left to right or circle decimal places in the number.

Learning Styles Students can:

 draw each item with a price.

 manipulate cards made from supermarket ads to organize items and prices before calculating.

 say multiplication problems aloud and count decimal places aloud.

Reteaching Activity

Provide hundredth grids or Visual 12 *Hundredths Model*. Students can use a grid to find decimal products, as shown below.

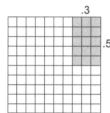

.5 × .3 = .15

Alternative Assessment

Students can use coins or Visual 15 *Money Models* for multiplication.

Example: Show why 15 × .05 = .75

Answer: Students should choose 15 nickels and find the total to be 75 cents.

10.8 Multiplying Decimals by 10, 100, 1,000
Student Edition page 214

Prerequisite Skill
- Multiplying whole numbers by 10, 100, and 1,000

Lesson Objective
- Multiply a decimal by 10, 100, and 1,000.

Cooperative Group Activity

Who Am I?
Materials: none

Procedure: Write .536 on the chalkboard. To the right of the number, write 53.6. Allow enough space between the two numbers for students to write "× 10," "× 100," or "× 1,000." Point first to the number on the left and then to the number on its right. Say, "I can turn the number on the left into the number on the right. Who am I?"

- A volunteer comes to the chalkboard and writes "× 100" between the two numbers. The volunteer counts the number of zeros in the power of 10 and counts the number of places the decimal point moves to change the number on the left to the number on the right. The class verifies that the number of zeros and the number of places that are moved are the same.

- The same volunteer writes a new pair of decimal numbers on the chalkboard and asks the "Who am I?" question.

- The activity proceeds in the same manner, using different volunteers each time.

Customizing the Activity for Individual Needs
ESL Be sure that students can differentiate between the words *right* and *left*. Have them point to different objects in the room that are to their right or to their left as they say the words. Then have them move their fingers different numbers of spaces to the right or left from a given number on a number line, saying the words aloud.

Learning Styles Students can:

 write down numbers on paper to use when it is their turn at the chalkboard.

 draw scoops with arrows under a number to show the movement of the decimal point.

 say aloud the number of places the decimal point moves and why. For example: "The decimal point moves 1 place to the right when the number is multiplied by 10 because 10 has 1 zero."

Enrichment Activity

Have students make posters comparing the pattern of multiplying by 10, 100, and 1,000 in whole numbers and in decimals. Encourage students to write about similarities and differences.

Alternative Assessment

Students can identify the product of a power of 10 and a decimal from a written set of choices.

Example: Which answer below shows .035 × 1,000?

.35   3.5   35   350   3,500   35,000

Answer: 35

ON-THE-JOB MATH
Bank Teller
Student Edition page 215

Lesson Objectives
- Apply knowledge of decimals to banking.
- Life Skill: Work with money.
- Communication Skill: Write a letter.
- Workplace Skill: Help customers with their banking needs.
- Workplace Skill: Understands deposit and withdrawal slips and account balances.

Activities

Write a Business Letter
Materials: pen and paper or computer processor
Procedure: Have students write a letter to a local bank to ask about their Professional Teller Training Program. Have them find out:
- how you can qualify.
- the date and time of the training.
- the location.
- what you need to bring.

Tools of the Trade
Materials: blank deposit and withdrawal slips, Visual 15 *Money Models*
Procedure: Have students work with partners to fill out actual deposit slips.
- One student writes amounts to be deposited on a deposit slip. The other student calculates the total, checks it, and then corrects any mistakes that occur. Reverse roles.
- One student fills out a withdrawal slip and gives it to his or her partner. The partner counts out the money shown on the slip.

Act It Out
Materials: none
Procedure: Present the following scenario for the students to role-play.

- Mr. Smith deposited a check for $250 yesterday. With the deposit, the balance in his account comes to $312.79. However, the check takes 2 days to clear.
- Mr. Smith wants to withdraw $100 from his account today. The teller must explain to Mr. Smith why he cannot withdraw $100 today.

Practice
Have students complete On-the-Job Math 133 *Bank Teller* from the Classroom Resource Binder.

Internet Connection
Have students use the following Web sites to find out about banking and bank tellers.
Sovereign Bank–interactive site about savings accounts
    www.kidsbank.com
Michigan Occupational Information Systems—tasks and work environment of bank tellers
    mois.org/scripts/031.HTM

10.9 Dividing Decimals by Whole Numbers
Student Edition pages 216–217

Prerequisite Skill
- Dividing whole numbers

Lesson Objectives
- Divide a decimal by a whole number.
- Life Skill: Compare unit costs.

Cooperative Group Activity

Finding Unit Prices
Materials: play money; paper; pencils; sale items, such as oranges and pens with price cards or pictures of items

Procedure: Organize students into groups of four or five. Prepare price cards such as "5 oranges for $2," "3 pens for $1." Set up one shopping station per group. Assign a cashier for each group.
- Each student may buy one of each item. Before paying, the student figures out the price of one item, or the *unit price*.
- The cashier must also figure out the unit price of each item.

- Students pay the cashier if the unit prices match. Students check their work if they do not agree with the cashier on the unit price.
- Groups move on to the next shopping station and repeat.

Customizing the Activity for Individual Needs

ESL Encourage students to use descriptive language when telling the part of a decimal that repeats. For example: "The 3 in the tenths place repeats for the decimal 1.3333...."

Learning Styles Students can:

 check to be sure every digit to the right of the decimal in the dividend has a digit above it in the quotient.

 draw decimal models on grid paper and then cut the models into equal parts so that they can be divided by whole numbers.

 say the steps aloud as they divide and listen to others as they do the same.

Enrichment Activity

Provide students with price signs such as the following: 4 for $1.99 or 50 cents each; 3 for $10 or $3.39 each. Have students figure out how much they save on the unit cost of each item by buying the larger quantity. Encourage students to create their own price signs to share with classmates.

Alternative Assessment

Students can correctly place zeros in the quotient and dividend when dividing a decimal by a whole number.

Example: When you divide .35 by 4, do you need a zero in the quotient? Where? Do you need zeros in the dividend? Why?

Answers: Yes, in the tenths place. Yes, to complete the division.

10.10 Dividing Decimals by Decimals
Student Edition pages 218–219

Prerequisite Skill
- Dividing whole numbers

Lesson Objectives
- Divide a decimal by a decimal.

- Life Skill: Find number of units sold and cost per unit.

Cooperative Group Activity

Find the Number of Units

Materials: index cards, paper, pencils

Procedure: Organize students into groups of four. Give each student an index card. Have students make price cards with unit costs and totals sales, such as "$.50 each; total cost $3.50." Collect the cards and shuffle them. Give each group member a price card.

- Using the information on their price cards, students figure out how many units were bought.
- Group members check each other's work and help each other to correct errors.

Customizing the Activity for Individual Needs

ESL To reinforce dividing with decimals, have pairs work together to make labeled flash cards showing how the decimal points should move to the right. One writes the problem as the other says the steps aloud.

Learning Styles Students can:

 create a poster to show the steps to follow when dividing decimals by decimals.

 draw scoops and arrows to show how the decimal point moves in a division problem.

 make a tape recording of the steps as they move the decimal point, add zeros, and do the division.

Reinforcement Activity

Write decimal numbers on slips of paper. Have students choose two slips to make up a division problem. Have students do the division. Then have pairs check each other's answers, correcting errors if necessary.

Alternative Assessment

Students can show their understanding of decimal division by matching division problems with their quotients.

Example: Match each division problem with its quotient.

1. .4)‾.032   a. 80
2. .4)‾.32   b. .08
3. .4)‾3.2   c. .8
4. .4)‾32   d. 8

Answers: 1. b; 2. c; 3. d; 4. a

10.11 Dividing Decimals by 10, 100, 1,000
Student Edition page 220

Prerequisite Skills
- Dividing decimals
- Dividing whole numbers by 10, 100, and 1,000

Lesson Objective
- Divide a decimal number by a power of 10.

Cooperative Group Activity

Who Am I?
Materials: none

Procedure: Write 78.2 on the chalkboard. To the right of the number, write 7.82. Allow enough space between the two numbers for students to write "÷ 10," "÷ 100," or "÷ 1,000." Point first to the number on the left and then to the number to its right. Say: "I can turn the number on the left into the number on the right. Who am I?"

- A volunteer comes to the board and writes "÷ 10" between the two numbers. The volunteer counts the number of zeros in the power of 10. The volunteer counts the number of places the decimal point moves to change the number on the left to the number on the right. The class verifies that the number of zeros and the number of places that are moved are the same.
- The same volunteer writes a new set of decimal numbers on the board and asks the "Who am I?" question.
- The activity proceeds in the same manner, using different volunteers each time.

Customizing the Activity for Individual Needs
ESL Have students create self-study cards to help them understand the direction for moving the decimal point to the *right* for *multiplication* and to the *left* for *division*.

Learning Styles Students can:

 highlight the number of zeros in the power of 10 and use the same color to draw scoops to the left.

 use a finger to trace the movement of the decimal point.

 say aloud the number of places the decimal point moves and why. For example: "The decimal point moves 1 place to the left when the number is divided by 10 because 10 has 1 zero."

Enrichment Activity
Have students make up posters comparing the pattern of dividing by 10, 100, and 1,000 in whole numbers and in decimals. Encourage students to write about similarities and differences.

Alternative Assessment
Students can identify the quotient of a decimal number and a power of 10 from a written set of choices.

Example: Which answer below is 3.5 × 1,000?

.0035   .035   .35   35   350   3,500

Answer: 3,500

USING YOUR CALCULATOR
Number Patterns
Student Edition page 221

Lesson Objectives
- Find patterns to solve problems.
- Life Skill: Compute with decimals on a calculator.

Activity
How Close Can You Get?
Materials: sign with numbers as shown below, calculators

Procedure: Have students work in pairs to play the game.
- A player looks at the directions and target number for each problem, and then chooses numbers from the sign to add or subtract.
- The sum or difference should be as close to the target number as possible.
- The player with the sum or difference that is closest to the target scores 1,000 points.

1. Add two numbers.
   Target: 1.5
2. Add two numbers.
   Target: 3.25
3. Add three numbers.
   Target: 15
4. Subtract two numbers.
   Target: .5
5. Subtract two numbers.
   Target: 5.5
6. Subtract two numbers.
   Target: 3.01

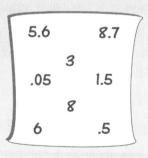

5.6   8.7
  3
.05   1.5
  8
6   .5

10.12 Problem Solving: Multi-Part Decimal Problems
Student Edition pages 222–223

Prerequisite Skill
- Solving two-part problems

Lesson Objectives
- Solve multi-part decimal word problems.
- Work backward to solve problems.

Cooperative Group Activity

Sports Fanatic
Materials: sports sections of newspapers; paper, pencils

Procedure: Assign students to groups. Distribute sports sections to each group. Have members find statistics for their favorite players.
- Each group creates five multi-part word problems using the statistics of each player.

  Example: Jamal Anderson—22.7 rushing yards; Dorsey Levins—25.2 rushing yards; Steve Young—26.3 passing yards. How many rushing yards do the players have combined? How many total yards do they have?

- The group calculates all the problems and records answers on paper. Then students highlight clue words and exchange them with another group to solve problems.

Customizing the Activity for Individual Needs
ESL To help students understand the words *sports statistics*, talk about which sports are popular in their countries of origin. Have them use statistics from these sports in the activity.

Learning Styles Students can:

 use a marker to highlight clue words in a word problem.

 use tools, such as Visual 22 *Problem Solving Steps*, to show the math process.

 read aloud the word problems and the steps to solve them.

Enrichment Activity
Have students work together to write parts of a word problem. Pairs can work together to combine the parts into multi-part problems. Groups exchange word problems and solve.

Alternative Assessment
Students can write the plan to solve a given problem.

Example: Susan bought bread that cost $1.59, milk that cost $2.29, and butter that cost $.89. How much is the change from $5?

Answer: Add the cost of each item. Subtract the sum from $5; the change is $.23.

10.13 Renaming Decimals as Fractions
Student Edition page 224

Prerequisite Skills
- Understanding decimal place value
- Reducing fractions to lowest terms

Lesson Objective
- Rename decimals as fractions in lowest terms.

Cooperative Group Activity

Concentration Part I
Materials: construction paper of two different colors cut into quarters to make cards, markers

Procedure: Organize students into groups of three or four. Give each group 18 cards of each color.
- Groups write decimal numbers on 18 cards of one color. Then they write an equivalent fraction for each of their decimal numbers on the other color cards. They shuffle the cards and turn them facedown in a 6 by 6 array.
- One student in each group turns over one card of the first color. The student then turns over a card of the second color. If the decimal number and fraction are equal, the cards match and are picked up. If not, they are turned over again and another student takes a turn.
- The game is over when all cards are picked up.
- Groups trade cards and play another new game.

Customizing the Activity for Individual Needs
ESL Help students understand which denominator to choose when renaming decimals as fractions. Students should read the decimal aloud while writing it in a place-value chart. When they have located the place value of the last digit, they use that place as the denominator of the fraction and simplify.

Learning Styles Students can:

 write the decimal in a place-value chart before renaming it as a fraction.

 tap a pencil as they count the number of places to the right of the decimal. Then they should write a "1" followed by that many zeros for the denominator.

 read the decimal number aloud. Then they should use the last place value named as the denominator.

Reteaching Activity

Provide students with copies of hundredths grids or Visual 12 *Hundredths Model*. Have students shade the grids to model a decimal number. Have them write the same number as a fraction, using the model as a guide, and then reduce to lowest terms.

Alternative Assessment

Students can write a decimal as a fraction by telling the steps to another student.

Example: Tell how to rename .008 as a fraction in lowest terms.

Sample answer: Because 8 is in the thousandths place, write .008 as $\frac{8}{1000}$. Then divide numerator and denominator by 8 to get $\frac{1}{125}$.

10.14 Renaming Fractions as Decimals
Student Edition page 225

Prerequisite Skill
- Dividing whole numbers

Lesson Objective
- Rename fractions as decimals.

Cooperative Group Activity

Concentration Part II

Materials: index cards of two different colors, markers

Procedure: Organize students into groups of three or four. Give each group 18 index cards of each color.

- Group members write fractions on 18 cards of one color. Then they write an equivalent decimal number for each of their fraction numbers on the other 18 cards. They shuffle the cards and turn them facedown in a 6-by-6 array.

- One student in each group turns over one card of the first color. The student then turns over a card of the second color. If the fraction and decimal number are equal, the cards match and are picked up. If not, they are turned over again and another student takes a turn.
- The game is over when all cards are picked up.
- The winner is the player with the most cards.
- Groups trade cards and play another game.

Customizing the Activity for Individual Needs

ESL To help students rename fractions as decimals, have them read the fraction aloud. Then have them tell the division problem that needs to be solved. *Example*: "Three-fourths means 3 divided by 4, so I write 3 ÷ 4 or $4\overline{)3}$."

Learning Styles Students can:

 write the fraction in higher terms as a number over 10, 100, or 1,000, and then rename that fraction as a decimal.

 write each part of the fraction on separate self-stick notes and then move the notes to form the division problem that needs to be solved.

 say the division aloud when determining whether the fraction and decimal are equivalent.

Reinforcement Activity

Have students create their own charts of fraction-decimal equivalents for halves, thirds, fourths, fifths, sixths, eighths, ninths, and tenths.

Alternative Assessment

Students can write a fraction as a decimal by telling the steps to another student.

Example: Tell how to rename $\frac{4}{5}$ as a decimal.

Sample answer: Divide 4 by 5. Add zeros to the dividend if needed. The quotient is .8.

10.15 Rounding Decimals
Student Edition pages 226–227

Prerequisite Skills
- Rounding whole numbers
- Using decimal place value

Lesson Objectives
- Round decimals.
- Life Skill: Round money to estimate amounts.

Cooperative Group Activity

Rounds of Rounding

Materials: index cards, pencils

Procedure: Organize students into groups of four. Give each student four index cards. Have them write four different decimal numbers to the thousandths place on separate cards. Collect the cards. Give each group an equal number of decimal index cards.

- For the first round, each group picks a card and rounds the number to the nearest hundredth.
- A volunteer from each group explains to the class how the number was rounded.
- For the second round, groups round the same number to the nearest tenth.
- Students repeat the activity until all the numbers have been rounded to the nearest hundredth and the nearest tenth.

Customizing the Activity for Individual Needs

ESL Give students copies of Visual 1 *Number Lines: Whole Numbers* or have them draw their own number lines. Show them how to label the blank number lines to help them determine which number a rounded decimal is closer to.

Learning Styles Students can:

highlight the digit in the rounding place. They then draw a line through it if the digit to its right is 5 or greater. They draw a circle around it if the digit to its right is less than 5.

write the digits of a decimal number on individual index cards. Then they can remove the cards to the right of the rounding digit when rounding down. They can replace the rounding digit with the next larger digit and remove the cards to the right when rounding up.

write and record a radio commercial that tells the rules for rounding decimals.

Reinforcement Activity

Give students the following problem and have them figure the problem in two ways: by rounding down to the nearest dollar and by rounding up to the nearest dollar. Ask why it is better to round up money amounts.

> Jannie has $20. She wants to buy three items costing $8.49, $4.49, and $7.49. Does she have enough money?

Alternative Assessment

Students can round a decimal to a place value by answering questions.

Example: Is 2.67 closer to 2.7 or 2.6 on the number line? *Answer:* 2.7

Closing the Chapter
Student Edition pages 228–229

Chapter Vocabulary
Review with students the Words to Know on page 201 of the Student Edition. Have students find the first use of each word in the chapter. Challenge them to write a sentence of their own for each word.

Have students copy and complete the Vocabulary Review on page 228 of the Student Edition. For more vocabulary practice, have them complete Vocabulary 119 *Decimals* from the Classroom Resource Binder.

Test Tips
Test tips with examples can be copied onto index cards. Labels for easy reference for this chapter may include: Reading and Writing; Comparing and Ordering; Adding and Subtracting; Multiplying and Dividing; Solving Problems.

Learning Objectives
Have students review CM6 *Goals and Self-Check*. They can check off the goal they have reached.

Note that each section of the quiz corresponds to a Learning Objective.

Group Activity
Summary: Students shop for a family of three for a week on a $125 budget.

Materials: grocery flyers and circulars; catalogs and brochures; calculators; Visual 16 *Blank Table*

Procedure: Encourage students to brainstorm a list of what they want before looking through the circulars. You might assign roles to group members, such as recorder, money counter, reporter, and researcher.

Assessment: Use the Group Activity Rubric on page xii of this guide. Fill in the rubric with the additional information below. For this project, students should have:

- selected healthy foods and necessary supplies.
- spent a total amount within the budgeted $125.

RELATED MATERIALS See the unit overview page for other Globe Fearon books that can be used to enrich and extend the material in this unit

Assessing the Chapter

Traditional Assessment
Chapter Quiz
The Chapter Quiz on pages 228–229 of the Student Edition can be used as either an open-book or closed-book test, or as homework. The quiz can be used to identify concepts in the chapter that students need to review and practice.

Chapter Tests
Use Chapter Test A Exercise 135 and Chapter Test B Exercise 136, *Decimals* from the Classroom Resource Binder to further assess mastery of chapter concepts.

Additional Resources
Use the Resource Planner on pages 116–117 of this guide to assign additional exercises from the Classroom Resource Binder and Workbook.

Alternative Assessment
Presentation
Have students explain the concept of adding decimals to the rest of the class, using hundredths grids or Visual 12 *Hundredths Model* and markers. Ask students to illustrate the concept and give two examples.

Option: Provide a transparency of Visual 12 and an overhead projector for students to use in their presentations.

Application
Have students find values of rolls or stacks of coins. For example: find the value of:
- 20 quarters
- 80 dimes
- 30 nickels
- 6 rolls of 50 pennies

Have students compare and order the values. Help them find how many of each kind of coin makes up the same value.

Planning the Chapter

Chapter 11 • Percents

Chapter at a Glance

SE page	Lesson	
230		Chapter Opener and Project
232	11.1	What Is a Percent?
233	11.2	Changing Percents to Decimals
234	11.3	Finding the Part
236	11.4	Sales Tax
238	11.5	Discounts
240	11.6	Commissions
242		Math in Your Life: Tipping
243	11.7	Changing Decimals to Percents
244	11.8	Changing Fractions to Percents
245		Using Your Calculator: The Percent Key
246	11.9	Finding the Percent
248	11.10	Finding the Percent Increase/Decrease
250	11.11	Finding the Whole
252	11.12	Finding the Original Price
254	11.13	Problem Solving: Finding the Part, Percent, or Whole
256		Chapter Review and Group Activity

Learning Objectives

- Write percents.
- Change among percents, decimals, and fractions.
- Find the part, the percent, and the whole in a problem.
- Find sales tax, discount, and commission.
- Find the percent increase or decrease and the original number.
- Solve problems about percents.
- Apply percents to find a tip.

Life Skills

- Find sale price and total cost, including sales tax.
- Calculate gross salary and tips.
- Find the percent of games won.
- Calculate the original price.
- Use the percent key on a calculator.
- Calculate savings account interest.
- Recognize a fair advertisement.

Communication Skills

- Read nutrition facts on food labels.
- Explain how to calculate a tip.
- Conduct an interview.

PREREQUISITE SKILLS
To assess mastery of prerequisite skills, use a selection of exercises from any of the program resources referenced below. The same resources can be used to provide remediation if necessary.

	Program Resources for Review		
Skills	Student Edition	Workbook	Classroom Resource Binder
Understanding decimals	Lessons 10.1, 10.2	Exercises 84, 85	Review 120
Multiplying decimals	Lesson 10.7	Exercise 90	Review 123 Practice 129
Adding and subtracting decimals	Lessons 10.5, 10.6	Exercises 88, 89	Review 121, 122 Practice 127, 128 Mixed Practice 132
Renaming fractions as decimals	Lesson 10.14	Exercise 97	
Dividing by a decimal	Lesson 10.10	Exercise 93	Review 125 Practice 130

Chapter 11 • 131

Diagnostic and Placement Guide

**Chapter 11
Percents**

The Percent Accuracy scores are based on the number of problems in a lesson that have been answered correctly.

Resource Planner

After diagnosing your students' needs, use the correlating Program Resources for reinforcement, reteaching, or enrichment. Additional activities for customizing lessons can be found in this guide.

KEY
Reteaching = ⤺   Reinforcement = ⬇   Enrichment = ⤻

Lessons		Percent Accuracy				Program Resources			
		50%	65%	80%	100%	⬇ Student Edition Extra Practice	⬇ Workbook Exercises	Teacher's Planning Guide	Classroom Resource Binder
11.1	What Is a Percent?	1	2	2	3	p. 433	99	⤺ p. 133	Visual 11 *Tenths Model* Visual 12 *Hundredths Model*
11.2	Changing Percents to Decimals	8	10	12	15	p. 433	100	⤺ p. 134	Visual 11 *Tenths Model* Visual 12 *Hundredths Model*
11.3	Finding the Part	11	14	18	22	p. 433	101	⤺ p. 135	⬇ Review 138 *Finding the Part*
11.4	Sales Tax	5	7	8	10	p. 434	102	⬇ p. 135	⬇ Review 139 *Sales Tax*
11.5	Discounts	5	7	8	10	p. 434	103	⤺ p. 136	⬇ Review 140 *Discounts*
11.6	Commissions	5	7	8	10	p. 435	104	⤺ p. 137	⬇ Review 141 *Commissions*
	Math in Your Life: Tipping	*Can be used for portfolio assessment.*						⤺ p. 137	⬇ Math in Your Life 147 *Tipping*
11.7	Changing Decimals to Percents	6	8	10	12	p. 435	105	⤺ p. 138	Visual 11 *Tenths Model* Visual 12 *Hundredths Model*
11.8	Changing Fractions to Percents	8	10	12	15	p. 435	106	⤺ p. 139	Visual 11 *Tenths Model* Visual 12 *Hundredths Model*
11.9	Finding the Percent	7	9	11	14	p. 436	107	⤺ p. 140	⬇ Review 142 *Finding the Percent*
11.10	Finding the Percent Increase/Decrease	5	7	8	10	p. 436	108	⤺ p. 140	⬇ Review 143 *Finding the Percent Increase/Decrease*
11.11	Finding the Whole	10	13	16	20	p. 436	109	⤺ p. 141	⬇ Review 144 *Finding the Whole*
11.12	Finding the Original Price	4	5	6	8	p. 437	110	⤺ p. 142	⬇ Review 145 *Finding the Original Price* ⤻ Challenge 148 *Get the Best Price*
11.13	Problem Solving: Finding the Part, Percent, or Whole	1	2	2	3	p. 437	111	⤺ p. 143	⬇ Mixed Practice 146 *Problem Solving: Finding the Part, Percent, or Whole*
Chapter 11 Review									
	Vocabulary Review	6	7.5 *(writing is worth 4 points)*	9.5	12			⬇ p. 144	⬇ Vocabulary 137
	Chapter Quiz	8	10	12	15			⬇ p. 144	⬇ Chapter Test A, B 149, 150

Customizing the Chapter

Opening the Chapter
Student Edition pages 230–231

Photo Activity
Procedure: Float an ice cube in a glass of water and have students estimate how much ice is above the water (11%). Ask students how they could represent a whole iceberg as a percent (100%). Help students use percents to find the weight above and below the surface for icebergs weighing 225,000 tons and 14,000 tons. Discuss other places they've seen percents used, such as in retailing and banking.

Answers: 225,000: 24,750 above; 200,250 below; 14,000: 1,540 above; 12,460 below

Words to Know
Review with students the Words to Know on page 231 of the Student Edition. Help students remember these terms by drawing a concept map with *percent* in the center and branches to the uses of percents, such as salaries, shopping, and nutrition.

The following words and definitions are covered in this chapter.

percent a part of a whole divided into 100 parts

sales tax a tax that is a percentage of the price of an item

discount amount that a price is reduced

sale price price of an item after the discount is subtracted

commission payment based on a percent of sales

base salary salary before adding commission

gross salary commission plus the base salary

percent increase percent more than an original number

percent decrease percent less than an original number

Nutrition Facts Project
Summary: Students collect nutrition facts labels, record the %RDA for fat, sodium, carbohydrates, and protein, and then calculate the number of servings needed to get 100% of each nutrient.

Materials: nutrition labels, paper, pencils

Procedure: You might have students begin by collecting and organizing their data for homework. They then can compare the fat, sodium, protein, and carbohydrate content of various foods in class.

They may complete the calculations during math class or as homework after completing Lesson 11.3.

Assessment: Use the Individual Activity Rubric on page xi of this guide. Fill in the rubric with the additional information below. For this project, students should have:
- collected labels and recorded %RDA of fat, sodium, carbohydrates, and protein.
- found the number of servings needed for 100% of each nutrient.

Learning Objectives
Review with students the Learning Objectives on page 231 of the Student Edition before starting the chapter. Students can use the list as a learning guide. Suggest they write the objectives in a journal or use CM6 *Goals and Self-Check* in the Classroom Resource Binder.

After each lesson, have students write an example of the skill they learned under the appropriate objective. Suggest that students use these notes as a learning guide to help them study for the chapter test.

11.1 What Is a Percent?
Student Edition page 232

Prerequisite Skill
- Understanding of decimals

Lesson Objective
- Write percents.

Words to Know
percent

Cooperative Group Activity

Percent Concentration
Materials: decimal grids (graph paper cut into 10-by-10 cards) or Visual 12 *Hundredths Model* reduced in size, colored pencils, index cards

Procedure: Have students work in pairs. Give each pair six decimal grids and six blank cards.
- One student shades any number of squares to model a percent. The other student writes the percent on a blank card. Students change roles and repeat the activity until all grids and cards have been filled in.
- Pairs shuffle the cards and turn them facedown in a 6-by-6 array. They take turns turning over two cards. If a percent matches its model, the player keeps the cards.

- Play continues until all cards are matched. The player with more cards wins. Have pairs trade decks and play again.

Customizing the Activity for Individual Needs

ESL To increase understanding of the word *percent*, write the word and the symbol on the chalkboard. Explain that *percent* comes from the Latin words *per centum*, meaning *of 100*. Brainstorm other words from the same Latin root, such as *cent, century,* and *centavo*.

Learning Styles Students can:

 use different colors to shade each column in a decimal grid to make counting easier.

 cut out portions of a decimal grid to model percents.

 count shaded squares out loud, saying *percent* after the total has been counted.

Enrichment Activity

Take a field trip to a local bank to find out interest rates on savings accounts. Encourage students to represent the percents on decimal grids or Visual 12 *Hundredths Model*. Discuss how to represent $\frac{1}{2}$% on a decimal grid. Have students report their findings to the class. The class can decide which savings account is the best choice.

Alternative Assessment

Students can explain what a percent means, given a model.

Example: Why does this model represent 26%?

Answer: 26 out of 100 squares are shaded.

11.2 Changing Percents to Decimals Student Edition page 233

Prerequisite Skill
- Understanding of decimals

Lesson Objective
- Change percents to decimals.

Cooperative Group Activity

Percent–Decimal Card Game

Materials: index cards; pencils; paper; spinner labeled 1, 2, and 3 or Visual 17 *Spinners*

Procedure: Organize students into teams of three or four. Give each student two index cards and each team a spinner.

- Each team member writes a percent on one side of each index card and then writes the decimal equivalent on the back of each card. Encourage students to include percents such as $4\frac{1}{2}$%, 16.5 %, and 138%.
- After checking their work, teams trade cards to play the game. They place the set of cards, percent side up, in the center of the table. One player spins the spinner to see the number of points for a correct answer. Then he or she chooses a card and writes the decimal equivalent on a sheet of paper. Another student checks the answer by turning over the card. Spinner points are scored if the answer is correct. Play is repeated with the next player.
- Players can help each other. The first team with 12 points wins.

Customizing the Activity for Individual Needs

ESL Compare the words *whole number, decimal,* and *percent* by writing 45, .45, and 45% on the chalkboard. Have students classify each number and then read each number: *forty-five, forty-five hundredths, forty-five percent*.

Learning Styles Students can:

 draw arrows under the decimal point when changing from a percent to a decimal.

 draw a decimal point on an index card and move it when changing from a percent to a decimal.

 read aloud the decimal and percent, and then tell how the two numbers are similar.

Reteaching Activity

Provide students with 100 small counters and a decimal grid or Visual 12 *Hundredths Model*. Write a percent between 1% and 100% on the chalkboard. Have students count out that number of counters. Tell students that each counter represents a penny. Have them write the amount of money in cents and as a decimal dollar; then they should place each counter on a square on the grid and write the percent as a decimal.

Alternative Assessment

Students can identify the decimal equal to a percent from a written choice.

Example: Which decimal represents 124%?
   124.   12.4   1.24   .124

Answer: 1.24

11.3 Finding the Part
Student Edition pages 234–235

Prerequisite Skill
- Multiplying decimals

Lesson Objectives
- Find a percent of a number.
- Communication Skill: Use information from a chart.

Cooperative Group Activity

Spinning Percents

Materials: whole number cards or Visual 4 *Whole Numbers to 100*, six-part spinner or Visual 17 *Spinners*, paper, pencils

Procedure: Organize students into groups of four. Provide each group with number cards and a spinner.
- Have one student in each group write the following percents on the spinner: 5%, 10%, 25%, 50%, 90%, 120%.
- One student picks a number from 1 to 100. Another student spins the spinner to determine the percent. The third student finds the percent of the number picked. The recorder writes the problem down and checks the calculation.
- Students change roles and repeat the activity for a round. A round is completed when the students have had an opportunity to find the percent of a number.

Customizing the Activity for Individual Needs
ESL   Remind students that in mathematics the word *of* means "to multiply." Then discuss the meaning of the word *poll*. Discuss examples of situations where polls are used, such as in politics or popularity contests.

Learning Styles   Students can:

- highlight the *whole* in a problem to distinguish it from *percent* and *part*.

 write the *whole*, *percent*, and *part* on separate index cards; then they can rearrange the cards to form the multiplication sentence.

 repeat aloud, "Part is Percent times Whole" to recall the steps.

Enrichment Activity

Have students answer the following questions based on the chart.

After-School Jobs	
Supermarkets	Video Stores
60 students	45 students
65% girls, 35% boys	20% girls, 80% boys

1. If the number of students who worked in supermarkets were 120 and the percents stayed the same, how many girls would be working there? How many boys?

2. If the percent of girls who worked at video stores doubled, how many girls would be working there? What would be the percent of boys? Why? How many boys would this be?

Answers: **1.** 78 girls; 42 boys **2.** 18 girls; 60%; 100% − 40% = 60%; 27 boys

Alternative Assessment

Students can calculate three different percents of the same number and compare the results.

Example: Calculate: 20% of 50 and 106% of 50. Compare each result to 50. Explain.

Answer: 10; 53; 10 is less than 50 because 20% is less than 100%. 53 is greater than 50 because 106% is greater than 100%.

11.4 Sales Tax
Student Edition pages 236–237

Prerequisite Skill
- Adding decimals

Lesson Objectives
- Find sales tax.
- Life Skill: Find the total cost, including sales tax.

Words to Know
sales tax

Cooperative Group Activity

Taxing Math

Materials: catalogs or flyers, paper, pencils

Procedure: Organize students into small groups. Give each group a catalog or sales flyer and explain that the members of the group have $100 to spend. They must buy a minimum of five items. Tell students the local sales tax.

- Group members work together to select items to buy. They record each item and its price. They add the prices, calculate the amount of sales tax, and then find the total cost.
- Students try different combinations of items to see how close they can come to $100 without going over. Groups share results with the class.

Customizing the Activity for Individual Needs

ESL Discuss the meaning of *tax*: *a charge or payment for the support of a government.* Brainstorm a list of different taxes that students may have heard of, such as income tax, sales tax, property tax, and import tax.

Learning Styles Students can:

 write an outline of the steps to follow when finding total cost, including sales tax.

 use money models to model purchases, sales tax, and total cost.

 explain to group members how to find the sales tax for their purchases.

Reinforcement Activity

Provide students with 100 tokens, chips, or pennies. Have students find various amounts of sales tax on $1. For example, 6% is 6 cents tax on one dollar. Then have them double that amount for the tax on $2, and so on. Have students model the results.

Alternative Assessment

Students can find the amount of sales tax and the total cost of an item using a calculator.

Example: A book costs $8.00. The sales tax is 5%. How much sales tax is charged? What is the total cost of the book?

Answers: $0.40; $8.40

11.5 Discounts
Student Edition pages 238–239

Prerequisite Skill
- Subtracting decimals

Lesson Objectives
- Find the amount of a discount.
- Life Skill: Find the sale price of an item.

Words to Know
discount, sale price

Cooperative Group Activity

What a Sale!

Materials: poster board, markers, ads or flyers from local businesses (grocery, hardware, clothing, appliance), pencils, paper

Procedure: Organize students into groups of three or four. Assign each group a different business. Give each group poster board and markers.

- Students design a poster for a sales event at their assigned store. Group members decide which items to put on sale and what percent off they want to offer.
- On a separate sheet of paper, group members calculate the sale prices for the items on their posters. Attach it to the back as a solution.
- *Extension*: Posters can be displayed around the room. Groups make a shopping list of different items and see how much they will save by buying the items on sale.

Customizing the Activity for Individual Needs

ESL Help students differentiate between *regular price* and *sale price*. Draw a price tag like the one below and explain each part.

Learning Styles Students can:

 use one color to write all amounts related to the discount and another color to write amounts related to the cost of an item.

use money models to model each problem.

 describe the multiplication problem that needs to be solved to find the discount.

Enrichment Activity

Have students look for advertised sales in newspapers and magazines. Ask them to determine the price of each item at the advertised discount. Then have them find the total cost of several items, including sales tax.

Alternative Assessment

Students can find the discount and sale price of an item using a calculator.

Example: A suitcase costs $58.00. It is on sale at 20% off. What is the discount? What is the sale price?

Answers: $11.60; $46.40

11.6 Commissions
Student Edition pages 240–241

Prerequisite Skill
- Adding decimals

Lesson Objectives
- Find the amount of a commission.
- Identify the base salary.
- Life Skill: Find gross salary.

Words to Know
commission, base salary, gross salary

Cooperative Group Activity

How Much Will You Make?

Materials: paper, pencils

Procedure: Organize students into groups of three.
- Group members choose a type of business. They decide on the weekly base salary and the commission rate for a salesperson who works at the business chosen.
- Students take turns role-playing being customers and salespeople. The customer names a price for the purchase of an item. Each salesperson keeps track of his or her sales and then finds his or her commission. After selling five items, the salesperson adds the base salary to the commission to find gross salary for a week.

Customizing the Activity for Individual Needs

ESL Discuss the meaning of *salary* by having students share job experiences involving money that they've earned. Relate the word *base* to familiar uses of the word, such as *at bottom* or *basic*. Explain that *base salary* is the *basic* money you earn.

Learning Styles Students can:

 label the numbers given in a problem as *percent*, *whole*, and *part*. Then they can write the multiplication problems.

 rearrange prepared index cards to form the multiplication problems.

 say the steps for finding commission and gross salary to themselves while solving.

Enrichment Activity

Tell students that they've been offered a job selling furniture that pays 9% commission plus a base salary of $15,000 a year. For this job, they would sell an average of $3,800 worth of furniture each week. They know they need to earn at least $32,000 a year to cover all their expenses. Ask students whether they think they should accept this job. Explain.

Answer: $3,800 × .09 = 342 × 52 weeks = $17,784 in commissions; $15,000 + $17,784 = $32,784 per year. Yes, they can accept this job. They will be able to cover their expenses.

Alternative Assessment

Students can find the amount of commission and the gross salary using a calculator.

Example: Suppose you earn a base salary of $250 a week. Your sales totaled $500. Your commission is 5% of sales. What is the amount of your commission? What is your gross pay?

Answers: $25; $275

MATH IN YOUR LIFE
Tipping
Student Edition page 242

Lesson Objectives
- Apply percents to find a tip.
- Life Skill: Calculate a tip quickly.
- Communication Skill: Interview a waiter or waitress about tipping.

(Continued on p. 138)

(Continued from p. 137)

Activities

Look It Up
Materials: none

Procedure: Have students make a list of professions in which tipping is used to show appreciation for service, such as:
- Waiter or waitress in a restaurant.
- Delivery person.
- Taxi driver.
- Parking garage attendant.

Suggest that students consider a 15% tip for good service and a 20% tip for exceptional service. Students can create a chart of the cost for each service and the amount of a tip based on the above guidelines. Students may consider reasons to tip or not to tip a person for service.

Conduct an Interview
Materials: tape recorder or notepad and pencil

Procedure: Encourage students to interview a waiter or waitress at a local restaurant. Ask them to find out:
- what he or she thinks is a fair tip percent.
- why a tip is important to his or her income.

Act It Out
Materials: menus, Visual 15 *Money Models*

Procedure: Organize students into groups of three or four. Give each group a menu and paper money. Assign one student to be the waiter or waitress. The other group members order from the menu. The waiter or waitress finds the total and presents the customers with the check. The customers pay the check and decide how much to leave for a tip. Have students role-play each situation:
- Service was excellent.
- Service was terrible.
- Customers leave too little a tip.
- Customers leave a large tip.

Practice

Have students complete Math in Your Life 147 *Tipping* from the Classroom Resource Binder.

Internet Connection

Search for local restaurants in your region at:
  Yahoo: Business and Economy
    http://dir.yahoo.com/Business_and_Economy/Companies/Food/Restaurants/By_Region/
Use a complete guide to tipping at:
  The Original Tipping Page
    http://www.tipping.org/tipping/tipping.html

11.7 Changing Decimals to Percents *Student Edition page 243*

Prerequisite Skill
- Changing percents to decimals

Lesson Objective
- Change a decimal to a percent.

Cooperative Group Activity

Decimal Change
Materials: 1-inch squares of cardboard or poster board, paper bags, pencils

Procedure: Form groups of three. Give each student four cardboard squares and a paper bag.
- Each group member writes the digits 0, 1, 2, and 3 on separate squares. Group members put all their squares in the paper bag.
- One member of the group picks three squares from the bag. Another group member writes the digits in the order chosen to form a three-digit number. (If 0 is drawn first, the recorder makes a two-digit number.)
- The third group member picks another square from the bag. This square tells students in front of which digit to place the decimal point. If 0 is drawn, the number is considered a whole number.
- All group members record the decimal number and change it to a percent. Students rotate roles and repeat the activity several times.

Customizing the Activity for Individual Needs

ESL Have partners practice reading decimals and percents. Students take turns reading aloud a decimal and an equivalent percent as the other student points to them on paper.

Learning Styles Students can:

 draw arrows under the decimal point when changing a decimal to a percent.

 use a finger to "move" the decimal point when changing a decimal to a percent.

 read aloud a decimal and its equivalent percent and tell how the two are similar.

Reteaching Activity

Provide students with Visual 12 *Hundredths Models*. Have them shade the grid to represent the decimals .05, .8, and 1. Then have them tell the percent of the grid that is shaded.

Alternative Assessment

Students can identify the percent equal to a decimal from a written set of choices.

Example: Which is .035 as a percent?
35%   3.5%   .35%   .035%

Answer: 3.5%

11.8 Changing Fractions to Percents *Student Edition page 244*

Prerequisite Skill
- Renaming fractions as decimals

Lesson Objective
- Change a fraction to a percent.

Cooperative Group Activity

Moving Math

Materials: self-stick notes, different-colored markers

Procedure: Divide the class into groups of three. For each group, use one marker to write the numerator of a fraction on a self-stick note. Use a different-colored marker to write the denominator on a different self-stick note. Stick the notes to the chalkboard to form fractions. To the right of each fraction write a large division sign: $\overline{)}$.

- One student from each group comes to the chalkboard and moves the self-stick notes to form a division problem that will change the fraction to a decimal.
- A second student from each group comes to the chalkboard to do the division.
- The third student writes the quotient as a percent.
- Students repeat the activity with different fractions for as long as appropriate.

Customizing the Activity for Individual Needs

ESL To increase understanding of *fraction to percent*, have pairs make quick reference cards showing the steps. Have them title the card *Fraction to Percent*. Under the title, pairs write examples, such as $\frac{3}{4} \rightarrow \frac{75}{100} \rightarrow 75\%$.

Learning Styles Students can:

 use different colors for the numerator and denominator as a reminder of the order of the numbers in the division problem.

write numerators and denominators of fractions on self-stick notes. Then they rearrange the notes to form the division problem needed to change the fraction to a decimal.

explain the steps to the group as they work at the chalkboard.

Enrichment Activity

Ask students to change the following mixed numbers into percents:

$1\frac{1}{2}, 8\frac{3}{4}, 3\frac{4}{5}$.

Have them explain the steps verbally as they work. Then have them use grid paper to illustrate each percent.

Answers: 150%, 875%, 380%

Alternative Assessment

Students can explain the steps needed to change a fraction to a percent.

Example: Change $\frac{4}{5}$ to a percent. What division problem would you use? When do you stop dividing? How do you change the decimal to a percent?

Answer: 80%; $5\overline{)4}$ (divide the denominator into the numerator); when the quotient stops or repeats move the decimal point two places to the right and add a percent sign.

USING YOUR CALCULATOR
The Percent Key
Student Edition page 245

Lesson Objective
- Life Skill: Use a calculator to find a percent of a whole number.

Activity

How Close Can You Get?

Materials: calculator with a percent key

Procedure: Have students work in pairs. The goal is to choose a number in Part B that will come as close as possible to the answer obtained in Part A.

- For Part A, the first player writes down a percent on a piece of paper. Then the student enters the percent into the calculator. The second player chooses a two- or three-digit number to be the whole. This player writes the percent problem and then finds the percent of the whole on the calculator.

(Continued on p. 140)

(Continued from p. 139)

- For Part B, the second player then chooses a different percent, writes it down, and enters it into the calculator. The first player chooses a number to be multiplied by the new percent. The product should be close to the number generated by player 2.

Example:

Player 1 chooses 25%.

    PRESS [2] [5] [%]   [.25]

Player 2 chooses 400.

    PRESS [×] [4] [0] [0] [=]   [100.]

Player 2 clears the display.

    PRESS [C]   [0.]

Player 2 chooses 50%.

    PRESS [5] [0] [%]   [0.5]

Player 1 chooses 200.

    PRESS [×] [2] [0] [0] [=]   [100.]

11.9 Finding the Percent
Student Edition pages 246–247

Prerequisite Skill
- Renaming a fraction as a percent

Lesson Objectives
- Find the percent given the whole and the part.
- Life Skill: Interpret a chart from a newspaper.

Cooperative Group Activity

Yes or No? Polls
Materials: paper, pencils, poster board, markers

Procedure: Pair students. Ask students to write a question for a poll they will take of their class, grade, or school. The question should be one that can be answered by *yes* or *no*, such as, "Should there be a dress code in school?"

- Pairs create posters showing the results of their poll. The poster should include the question asked, the number of students polled, the percent answering yes, and the percent answering no. Students should check that the two percents add up to 100%.
- Pairs place their posters on a bulletin board titled *Polling and You!*

Customizing the Activity for Individual Needs

ESL Help students relate the terms *part*, *whole*, and *percent* to wording in different examples. Have them box the whole and circle the part. They can make a fraction using part/whole and then change the fraction to a decimal and the decimal to a percent. Have them highlight the percent.

Learning Styles Students can:

 use different colors for the part and whole. They then can use the same color codes when setting up the fractions and division problem.

 write the part and the whole on separate self-stick notes. They then arrange the notes to form a fraction, noting which is the divisor and which is the dividend.

 say aloud the following after finding the percent: "[part] is ____ percent of [whole]."

Enrichment Activity

Provide students with a week's attendance records for the class. Ask them to find what percent of students were present or absent each day for that week.

Alternative Assessment

Students can find the percent using a calculator.

Example: What percent of 40 is 24?

Answer: 60%

11.10 Finding the Percent Increase/Decrease
Student Edition pages 248–249

Prerequisite Skill
- Finding a percent

Lesson Objective
- Find the percent increase/decrease.

Words to Know
percent increase, percent decrease

Cooperative Group Activity

It's More or Less!
Materials: index cards, paper, pencils

Procedure: Organize students into small groups. Give each group 10 index cards.

140 • Chapter 11

- Students create two sets of number cards. On one side of five index cards, they write the words "Original Price." On the other side, they write the following prices on separate cards: $50, $60, $100, $120, and $400.
- On one side of the remaining five cards, they write the words "New Price." On the other side, they write: $30, $80, $140, $240, $300.
- Students make two piles of cards with the words "Original Price" and "New Price" facing up. Students take turns choosing one card from each pile. They then find the percent increase or decrease. Problems and solutions are recorded on a sheet of paper.
- Group members check each other's work. Shuffle the cards and repeat.

Variation: Have teams compete by awarding points for correct problems.

Customizing the Activity for Individual Needs

ESL To help reinforce the meanings of *increase* and *decrease*, demonstrate *more* and *less* of an amount.

Learning Styles Students can:

circle the original number in the word problem, subtraction problem, and fraction.

make a quick-reference index card containing an outline of the steps to follow.

say the steps used to solve the problem while writing the steps.

Enrichment Activity

Have students solve the following problem:

A store normally sells a jacket for $100. It decided to increase this price by 20%. A few months later, it decreased the new price by 20%. Is the jacket back to its original price of $100? Explain your reasoning. Support it with your calculations.

Answer: The new jacket is $96. The price is decreased by 20% of the new price, which is greater than $100.

Alternative Assessment

Students can explain the steps for finding the percent increase or decrease and then use a calculator to do the math.

Example: The price of a bike was $90 in December and $72 in January. Was there a percent increase or decrease? How do you know? What was the percent increase or decrease? How did you find it?

Sample Answer: The percent decrease was 20%. Since 72 is less than 90, it is a percent decrease. First subtract 72 from 90. Divide the result by 90, the original number. Move the decimal point in the quotient two places to the right.

11.11 Finding the Whole
Student Edition pages 250–251

Prerequisite Skills
- Changing a percent to a decimal
- Dividing by a decimal

Lesson Objectives
- Find the whole given the percent and part.
- Life Skill: Find interest on a savings account.

Cooperative Group Activity

Interest-Rate Quick Draw

Materials: index cards, paper, pencils, savings-account brochures from local banks showing possible interest rates

Procedure: Organize students into small groups. Give each group 10 index cards and some bank brochures.

- Groups look at brochures and identify five possible interest rates. On one side of five index cards, they write the bank name, account type, and the words *Interest Rate*. On the other side, they write the actual interest rate.
- On one side of the remaining five cards, they write *Interest Earned*. On the other side, they write the following numbers: $5, $10, $100, $500, and $1,000.
- Groups make two piles of cards with the words *Interest Rate* and *Interest Earned* facing up. Students take turns choosing one card from each pile. They then calculate how much savings or principal they would need in order to earn the amount of interest at the chosen rate. Problems and solutions are recorded on a sheet of paper.
- Group members check each other's work. Shuffle the cards and repeat.

Variation: Have teams compete by awarding points for correct problems.

Customizing the Activity for Individual Needs

ESL Help students become familiar with the mathematical use of the words *interest* and *principal*. A bank pays money (interest) on the amount (principal) in a savings account.

Learning Styles Students can:

 make a labeled reference card to show the steps needed to find the whole given the percent and the part.

 use a finger to show moving the decimal point when changing the percent to a decimal and when dividing by the decimal.

 form a question for the class to answer based on the cards chosen. For example, "How much savings would you need to earn $450 in interest at a rate of 6%?" *Answer:* $7,500

Reteaching Activity

Give students copies of Visual 11 *Tenths Model* and markers. Write, "24 is 30% of what number?" on the chalkboard. Have students color three sections of the tenths model to represent 30%. Ask students how to divide 24 equally among the three sections. Have them write *8* on each section. Tell students that each 10% section equals 8, so the whole (100%) must be $8 \times 10 = 80$. Therefore, 24 is 30% of 80. Repeat for problems such as "80 is 40% of what number?" and so on.

Alternative Assessment

Students can give step-by-step directions explaining how to find the whole, given a part and the percent.

Example: Give the steps you would use to solve this problem: 30 is 75% of what number?

Sample answer: Step 1: change 75% to .75; Step 2: divide 30 by .75; Step 3: the quotient, 40, is the whole.

11.12 Finding the Original Price
Student Edition pages 252–253

Prerequisite Skill
- Dividing by a decimal

Lesson Objective
- Find the original price.

Cooperative Group Activity

The Price Is Right
Materials: various shopping catalogs, self-stick notes, paper, pencils, scissors

Procedure: Organize students into small groups. Give each group catalogs, scissors, and self-stick notes.

- Groups go through catalogs and select and cut out at least one item for each member. For each item, students write the original price on a self-stick note and stick it to the back of the picture of the item.
- Then students "mark up" or "mark down" the item using a percent increase or decrease. The new price and percent increase or decrease is written on a self-stick note on the front of the picture.
- Groups exchange sale items and try to solve for the original price. Answers can be checked by looking on the reverse side.

Customizing the Activity for Individual Needs

ESL Help students correlate the terms *percent*, *original price*, and *new price* to *percent*, *whole*, and *part*. Have student pairs work together. One student highlights the three parts of the problem. The other student labels the highlighted parts as *percent*, *part*, or *whole*.

Learning Styles Students can:

 use colors to identify a price markup as an increase or price mark-down as a decrease.

 fill in price tags and put them on real-life items to model the problems and help decide whether the new price is an increase or decrease.

 record each of the steps for finding the original price on an audiotape recorder as they solve a problem.

Enrichment Activity

Have students solve the following problem:

The same CD player is on sale at two different stores. In Store A, it is on sale for $112, which is 20% off the original price. In Store B, it is on sale for $105, which is 30% off the original price. Which store has the better price for the CD player when it is not on sale?

Answer: Store A. The original price is $140 in Store A and $150 in Store B.

Alternative Assessment

Students can explain how to find the original price of an item.

Example: Find the original price of a pair of shoes if they are on sale for 30% off that price and cost $35. What percent of the original price is the new price? What percent problem would you solve? What division problem would you solve? What was the original price?

Answer: 100% − 30% = 70%; 70% of what number is 35; 35 ÷ .70; $50

11.13 Problem Solving: Finding the Part, Percent, or Whole
Student Edition pages 254–255

Prerequisite Skill
- Multiplying and dividing with percents

Lesson Objectives
- Solve word problems involving percent.
- Use sentence diagrams to help solve problems.

Cooperative Group Activity

Heads Together: Part, Percent, or Whole?

Materials: paper, pencils

Procedure: Organize the class into three groups. Assign each group one of the following types of problems: (a) Find the part, given the percent and the whole. (b) Find the percent, given the whole and the part. (c) Find the whole, given the percent and the part.

- Each student writes a word problem based on the group's assigned problem type.
- The problems are placed in separate piles in a central location. Each group picks one of each kind of problem. Group members work together to solve and check each problem.

Variation: Announce a problem type and randomly choose a group member to state a problem and explain the solution. A correct response earns the team an award.

Customizing the Activity for Individual Needs

ESL Help students identify key words that tell them which kind of problem to solve. Have them highlight and label the numbers in the problem as *percent*, *whole*, or *part*. Then they should choose from these number sentences to solve it.

whole = part ÷ percent
part = percent × whole
percent = part ÷ whole

Learning Styles Students can:

 make a reference card for each type of percent problem, and they will know how to solve it.

 manipulate numbers and symbols on self-stick notes to form the division or multiplication number sentence necessary to solve the problem.

 try to solve a problem while listening to an audiotape that explains each step for solving that kind of problem.

Reteaching Activity

Write the following on the chalkboard:

_____ % of _____ is _____
  percent   whole   part

Then write a percent word problem on the chalkboard using self-stick notes for the given numbers. Have a student take one of the numbers from the problem and stick it onto the appropriate part of the outline. Repeat for the other numbers. The student should write a question mark in the outline for the missing data. Then have the students solve the problem.

Alternative Assessment

Students can show their ability to solve percent word problems by finding any errors in the solution to a problem.

Example: Find and correct any errors in the solution to this problem: There are 50 ducks in a pond. Of the total, 18 are mallards. What percent of the ducks in the pond are mallards?

Solution: 50 ÷ 18 = 2.78; 278% are mallards

Answer: There cannot be more than 100% mallards. The problem that should be solved is 18 ÷ 50 = .36. 36% are mallards.

Closing the Chapter
Student Edition pages 256–257

Chapter Vocabulary
Review with students the Words to Know on page 231 of the Student Edition. Then have students quiz each other in pairs.

Have students copy and complete the Vocabulary Review questions on page 256 of the Student Edition.

For more vocabulary practice, use Vocabulary 137 *Percents* from the Classroom Resource Binder.

Test Tips
In pairs, students can take turns showing how to solve a problem by using one of the test tips.

Learning Objectives
Have students review CM6 *Goals and Self-Check* from the Classroom Resource Binder. They can check off the goal they have reached. Note that each section of the quiz corresponds to a Learning Objective.

Group Activity
Summary: Students find the percent of discount, original price, and sale price for three items in an advertisement.

Materials: advertisements, pencils, paper

Procedure: Have students organize their information into a chart or make posters for each item on construction paper.

Assessment: Use the Group Activity Rubric on page *xii* of this guide. Fill in the rubric with the additional information below. For this activity, students should have:
- found and recorded the percent discount, original price, and sale price.
- chosen three items and decided which store had the fairest advertisement.

RELATED MATERIALS See the unit overview page for other Globe Fearon books that can be used to enrich and extend the material in this unit.

Assessing the Chapter

Traditional Assessment
Chapter Quiz
The Chapter Quiz on pages 256–257 of the Student Edition can be used as either an open- or closed-book test, or as homework. The quiz can be used to identify concepts in the chapter that students need to review and practice.

Chapter Tests
Use Chapter Test A Exercise 149 and Chapter Test B Exercise 150 *Percents* from the Classroom Resource Binder to further assess mastery of chapter concepts.

Additional Resources
Use the Resource Planner on page 132 to assign additional exercises from the Classroom Resource Binder and Workbook.

Alternative Assessment
Performance-Based
Write the number 350. Have students respond to the following:
- What is 12% of this number?
- What is a 20% decrease in this number?
- What percent of this number is 140?
- Put a decimal between the first and second digits. Write it as a fraction and a percent.
- What is the percent increase or decrease if this number is now $280?

Real-Life Connection
Have students find percents in newspapers and magazines and then:
- write the percent as a fraction and a decimal.
- find that percent of 40.
- write one problem that can be solved using the percent found and the number 2,000. Then solve the problem.

Planning the Chapter

Chapter 12 • Ratios and Proportions

Chapter at a Glance

SE page	Lesson	
258		Chapter Opener and Project
260	12.1	What Is a Ratio?
262	12.2	What Is a Proportion?
264	12.3	Solving Proportions
265		Using Your Calculator: Solving Proportions
266	12.4	Multiple Unit Pricing
268	12.5	Scale Drawings
269		On-The-Job Math: Courier
270	12.6	Problem Solving: Using Proportions
272		Chapter Review and Group Activity
274		Unit Review

Learning Objectives
- Write ratios.
- Solve proportions.
- Find multiple unit prices.
- Use scale drawings to find actual sizes.
- Solve problems using proportions.
- Apply using map scales to find distances.

Life Skills
- Use ratios in recipes.
- Use a calculator to solve proportions.
- Make a scale drawing.
- Draw pictures to solve problems.

Communication Skills
- Read supermarket ads.
- Draw and label a scale drawing.
- Interpret maps and scale drawings.
- Write questions.
- Explain how to find the actual distance between two points on a map.

Workplace Skills
- Calculate distances along a delivery route.
- Map a delivery route.

PREREQUISITE SKILLS
To assess mastery of prerequisite skills, use a selection of exercises from any of the program resources referenced below. The same resources can be used to provide remediation if necessary.

Skills	Program Resources for Review		
	Student Edition	Workbook	Classroom Resource Binder
Multiplying whole numbers	Lessons 4.1, 4.2	Exercise 26, 27	Review 36–38 Practice 39, 40, 42
Reducing fractions to lowest terms	Lesson 7.3	Exercise 56	Review 78
Changing whole and mixed numbers to fractions	Lesson 7.9	Exercise 62	Practice 82
Changing fractions to mixed numbers	Lesson 7.8	Exercise 61	Practice 82

Diagnostic and Placement Guide

**Chapter 12
Ratios and Proportions**

The Percent Accuracy scores are based on the number of problems in a lesson that have been answered correctly.

Resource Planner

After diagnosing your students' needs, use the correlating Program Resources for reinforcement, reteaching, or enrichment. Additional activities for customizing lessons can be found in this guide.

KEY
Reteaching = ⤺   Reinforcement = ⬇   Enrichment = ⤻

Lessons		Percent Accuracy				Program Resources			
		50%	65%	80%	100%	⬇ Student Edition • Extra Practice	⬇ Workbook • Exercises	Teacher's Planning Guide	Classroom Resource Binder
12.1	What Is a Ratio?	2	3	3	4	p. 438,	112	⤺ p. 147	⬇ Review 152 *Ratio and Proportion*
12.2	What Is a Proportion?	9	12	14	18		113	⤺ p. 148	⬇ Review 152 *Ratio and Proportion*
12.3	Solving Proportions	6	8	10	12	p. 438	114	⤺ p. 149	⬇ Review 153 *Solving Proportions*
12.4	Multiple Unit Pricing	3	4	5	6	p. 438	115	⤺ p. 150	⬇ Review 154 *Multiple Unit Pricing*
12.5	Scale Drawings	2	3	3	4	p. 438	116	⤺ p. 150	⬇ Practice 155 *Scale Drawings* ⤺ Challenge 158 *Drawing Plans*
	On-The-Job Math: Courier	*Can be used for portfolio assessment.*						p. 151	⤺ On-The-Job Math 157 *Courier*
12.6	Problem Solving: Using Proportions	1	2	2	3	p. 438	117	⤺ p. 152	⬇ Practice 156 *Problem Solving: Using Proportions*
Chapter 9 Review									
	Vocabulary Review	6.5	7	8.5	11			p. 153	⬇ Vocabulary 151
		(writing is worth 4 points)							
	Chapter Quiz	7	9	11	14			p. 153	⬇ Chapter Test A, B 159, 160
Unit 3 Review		5	6.5	8	10			p. 153	⬇ Unit Test pp. T11–12
		(critical thinking is worth 4 points)							

Customizing the Chapter

Opening the Chapter
Student Edition pages 258–259

Photo Activity
Procedure: With students, discuss scale models. They may want to bring in scale models of cars, planes, rockets, trains, or furniture. The scale for these items is often found on the packaging. Ask: *How is a model the same as the actual object? How is it different? What determines the size of the model?*

Answers: A model is the same shape as the actual object; a model is either smaller or larger than the actual object; the scale determines the size of the model.

Point out that builders sometimes use scale models to show clients what a building will look like once it is constructed.

Words to Know
Review with the students the Words to Know on page 259 of the Student Edition. Help students remember these terms by having them give an example of each one along with its definition.

The following words and definitions are covered in this chapter.

ratio a comparison of one amount with another

proportion a statement that two ratios are equal

cross products the results of cross multiplying

rate a comparison of two amounts with different units of measure

multiple unit pricing the cost of a set of items

scale drawing a picture that shows the proportional size of actual objects

scale a ratio that compares the size of a drawing with the size of the original object

Scale Drawing Project
Summary: Students make a scale drawing of their bedroom on graph paper.

Materials: graph paper, tape measures, pencils

Procedure: Students can begin the project at home by measuring the dimensions of their bedrooms and the furniture pieces. Assist students in choosing appropriate scales for their drawings. Students can complete the scale drawing and label the actual measurements in class or for homework after completing Lesson 12.5.

Assessment: Use the Individual Activity Rubric on page xi of this guide. Fill in the rubric with the additional information below. For this project, students should have:
- a scale comparing inches to feet.
- an accurate scale drawing of their rooms.

Learning Objectives
Review the Learning Objectives on page 259 of the Student Edition before starting the chapter. Students can use the list of objectives as a learning guide. Suggest that they write the objectives in a journal or use CM6 *Goals and Self-Check* of the Classroom Resource Binder.

After each lesson, have students write an example of the skill they learned under the appropriate objective. Suggest that students use these notes as a learning guide to help them study for the chapter test.

12.1 What Is a Ratio?
Student Edition pages 260–261

Prerequisite Skill
- Reducing fractions to lowest terms

Lesson Objectives
- Write ratios to compare two quantities.
- Life Skill: Use ratios in recipes.

Word to Know
ratio

Cooperative Group Activity

Dictionary Ratios
Materials: dictionaries, paper, pencils

Procedure: Organize students into pairs. Give each pair a dictionary.
- One student chooses a word in the dictionary and writes the word on a sheet of paper.
- The partner writes six possible ratios for the word, comparing the number of: (1) vowels to all letters, (2) all letters to vowels, (3) consonants to all letters, (4) all letters to consonants, (5) vowels to consonants, and (6) consonants to vowels. Ratios should be expressed in lowest terms.

- Partners reverse roles. They repeat the activity until the letters in 10 words have been compared.

Customizing the Activity for Individual Needs

ESL Be sure that students know the difference between *consonants* and *vowels*. Have them write the letters of the alphabet, then circle all the vowels for reference.

Learning Styles Students can:

 use different colors to highlight or circle each quantity being compared and use the same colors to write the ratio.

 use two-color counters to model ratios.

 read aloud each ratio. For example, "The ratio of vowels to the total number of letters is ___ to ___."

Enrichment Activity

Provide students with sports sections from newspapers. Have them find statistics that can be expressed as ratios, such as wins to losses. Have students write the ratios numerically. Then have them explain the ratios to the class.

Alternative Assessment

Students can draw a diagram to illustrate a ratio.

Example: Use shaded circles to show what the ratio 2:3 could mean.

Answer: Students can shade two out of five circles to compare two shaded circles to three unshaded circles or shade two out of three circles to compare two shaded circles to all circles.

 or

12.2 What Is a Proportion?
Student Edition pages 262–263

Prerequisite Skills
- Writing ratios
- Multiplying whole numbers
- Comparing whole numbers

Lesson Objective
- Identify proportions.

Words to Know
- proportion, cross products

Cooperative Group Activity

Proportions Game

Materials: paper, pencils

Procedure: Divide the class into pairs of students. Write the following ratios in fraction form on the chalkboard: $\frac{4}{5}, \frac{9}{18}, \frac{12}{15}, \frac{12}{24}, \frac{6}{12}$.

- Each student chooses two ratios from the chalkboard that may form a proportion.
- Have students write the ratios and use cross products to decide if the pair of ratios does form a proportion. Partners check each other's work.
- Students who identify the four proportions are winners: $\frac{4}{5} = \frac{12}{15}$, $\frac{9}{18} = \frac{12}{24}$, $\frac{9}{18} = \frac{6}{12}$, and $\frac{6}{12} = \frac{12}{24}$.

Customizing the Activity for Individual Needs

ESL Be sure that students can differentiate between the words *ratio* and *proportion*.

Learning Styles Students can:

 draw and shade pairs of geometric shapes to represent proportions.

 shade and cut out pairs of geometric shapes that represent proportions. Then place one shaded part on top of the other to see if they coincide to test for proportions.

 orally explain which ratios form proportions, and why.

Reinforcement Activity

Have one student write a ratio and the other student use the ratio to write a proportion. Students can use cross products to verify proportions.

Alternative Assessment

Students can recognize proportions given geometric representations of ratios.

Example: Which two ratios form a proportion?

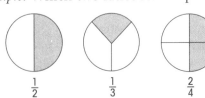

Answer: $\frac{1}{2}$ and $\frac{2}{4}$ form a proportion because $\frac{1}{2} = \frac{2}{4}$.

12.3 Solving Proportions
Student Edition page 264

Prerequisite Skills
- Multiplying whole numbers
- Dividing whole numbers
- Comparing whole numbers

Lesson Objective
- Find the missing number in a proportion.

Cooperative Group Activity

And the Missing Number Is . . .

Materials: paper, pencils, number cubes

Procedure: Assign students to groups of four. Give each group paper, pencils, and three number cubes.
- One student throws the three number cubes.
- Then each student in the group uses the three numbers to write a proportion with a missing term. The missing term should be in a different place for each proportion. For example, if the numbers 2, 5, and 3 are thrown, possible proportions are:

$\frac{2}{5} = \frac{3}{?}$; $\frac{2}{5} = \frac{?}{3}$; $\frac{2}{?} = \frac{5}{3}$; $\frac{?}{2} = \frac{5}{3}$.

- Students solve the proportion they write. They then use cross products to check.
- Students receive 1 point for each correct solution. Teams can compete for scores.

Customizing the Activity for Individual Needs

ESL Demonstrate how to solve a proportion on the chalkboard without using any words. Then have partners demonstrate the process to each other. Together, have them say and then write the steps in their own words.

Learning Styles Students can:

 draw arrows to show cross products and write out a check to make sure the cross products are equal.

 use number cards to show the computation used to solve a proportion.

 say the steps aloud while a partner follows them.

Enrichment Activity

Have students use a given number as the cross product of a proportion. Announce a number to the class. Have students write two proportions with that cross product. For example, for a cross product of 63, students can write:
$\frac{1}{7} = \frac{9}{63}$ and $\frac{3}{7} = \frac{9}{21}$.

Alternative Assessment

Students can explain how to solve a proportion.

Example: How would you find the missing number in this proportion? $\frac{10}{35} = \frac{30}{?}$

Answer: Multiply 30 × 35 to find the cross product, 1,050; divide by 10 to find the missing number; 105.

USING YOUR CALCULATOR
Solving Proportions
Student Edition page 265

Lesson Objectives
- Life Skill: Use a calculator to solve proportions.
- Life Skill: Use a calculator to check answers.

Activities

Stump Your Partner

Materials: paper, pencils, calculators

Procedure: Have students work in pairs.
- Each student writes a proportion with one missing number. The three numbers in the proportion can be randomly chosen.
- Students exchange papers and solve each other's proportion using a calculator.
- Students exchange papers again and check each other's work by calculating the cross products.
- Each correct answer is worth 5 points. The first partner to reach 30 points wins.

Missing Numbers

Materials: paper, pencils, calculators

Procedure: Have students work in groups of four. Illustrate the following proportions on the chalkboard.

$\frac{5}{?} = \frac{15}{27}$ $\frac{8}{6} = \frac{36}{?}$

$\frac{?}{5} = \frac{16}{40}$ $\frac{?}{16} = \frac{9}{8}$

$\frac{14}{?} = \frac{28}{32}$ $\frac{8}{12} = \frac{?}{30}$

(Continued on p. 150)

(Continued from p. 149)

- Each of three students chooses a different pair of proportions to solve using a calculator.
- The fourth student checks each solution by calculating the cross products.
- Each correct answer is worth 5 points.
- Students take turns being the checker.

12.4 Multiple Unit Pricing
Student Edition pages 266–267

Prerequisite Skill
- Solving proportions

Lesson Objectives
- Life Skill: Use proportions to find the price of any number of the same items given the multiple unit price.
- Life Skill: Use proportions to find the number of items that can be bought for a certain amount of money given the multiple unit price.

Words to Know
rate, multiple unit pricing

Cooperative Group Activity

Hunting for Bargains
Materials: supermarket ads from newspapers or fliers, paper, pencils

Procedure: Organize students into groups of three. Give each group a different supermarket ad or flier.
- One group member chooses an advertisement that contains multiple unit pricing. This ad is the basis for the rest of the activity.
- A second group member writes a problem asking for the cost of a certain number of the items in the ad.
- The third group member writes a problem asking for the number of items that can be bought for a certain dollar amount.
- All members solve each of the problems. They then compare answers.
- Students can rotate roles and repeat the activity, or groups can exchange problems to solve for further practice.

Customizing the Activity for Individual Needs

ESL Help students understand the terms *rate* and *multiple unit pricing* by giving examples such as, "If 1 pound of apples cost 49¢ then 5 pounds cost $2.45." Have students cut out examples of multiple unit pricing from supermarket ads, inserting the words *if. . .*, *then. . .*.

Learning Styles Students can:

 highlight the item amount or dollar amount in each ratio to help set up the correct proportion for solving.

 use money models and manipulatives to model multiple unit pricing problems.

 say aloud the rate and proportions needed to solve each problem.

Reteaching Activity

Provide students with common classroom objects, such as scissors, chalk, erasers, and pencils. Create a price tag for sets of objects that involve multiple unit pricing. An example might be "6 pencils for $1.50." Have students find the cost of nine pencils. Then have them find how many pencils they can buy with $5.

Answers: $2.25; 20 pencils

Alternative Assessment

Given multiple unit pricing, students can model how to find the price of twice the number of items using money models and manipulatives.

Example: Socks are priced at 3 pairs for $7. How much will 6 pairs of socks cost?

Answer: $14

12.5 Scale Drawings
Student Edition page 268

Prerequisite Skills
- Writing rates
- Solving proportions

Lesson Objectives
- Communication Skill: Interpret a scale drawing.
- Life Skill: Find the actual length of an object from a scale drawing.

Words to Know
scale drawing, scale

Cooperative Group Activity

Will It Fit?
Materials: furniture catalogs, paper, pencils

Procedure: Organize the class into groups of three students. Give each group a furniture catalog.

- Each group creates a list of furnishings from the catalog for the living room and bedroom shown in the scale drawing on page 268 of the Student Edition.
- Students find the actual size of the rooms. They then check the sizes of the furniture to make sure the furniture will fit. You may need to assist groups in determining fit.
- Students can revise choices as necessary to find furniture that will fit in the rooms.

Customizing the Activity for Individual Needs
ESL To increase comprehension of the word *scale*, ask students for different meanings of the word. Discuss the meaning of scale used in the text. Then discuss how this differs from a scale used to weigh things or from fish scales.

Learning Styles Students can:

 write the steps for finding the actual size of an object given a scale drawing of the object.

 use a ruler to measure the actual size of an object to help interpret a scale drawing.

 describe aloud the steps for finding the actual size of an object given a scale drawing.

Enrichment Activity
Have students make a scale drawing of the ground floor of their school. Provide them with tape measures and graph paper. Have volunteers measure the lengths of each side of the building. Help students select an appropriate scale. Have them write proportions to find the scale length of each side of the building. Then have them use their answers to draw the school on graph paper.

Alternative Assessment
Students can explain how to find the actual size of an object given a scale drawing of the object.

Example: How would you find the actual length of a room that is 6 inches long on a drawing whose scale is 1 inch equals 4 feet?

Answer: Set up the proportion $\frac{1 \text{ in.}}{4 \text{ ft}} = \frac{6}{?} \text{ ft}$
Find the cross product, 24, and divide by the remaining number, 1. The actual length is 24 feet.

ON-THE-JOB MATH
Courier
Student Edition page 269

Lesson Objectives
- Apply map scales to finding actual distances.
- Workplace Skill: Calculate distances along a delivery route.
- Workplace Skill: Map a route.
- Communication Skill: Explain how to find the actual distance between two points on a map.
- Communication Skill: Write questions to ask a speaker.

Activities

Tools of the Trade
Materials: town maps, pencils

Procedure: Divide the class into groups of three students.
- Using a town map, each group creates a delivery route within the town with at least fives places for delivery.
- Then, groups use the map scale to find the actual distance between each delivery.

Ask a Speaker
Materials: paper, pencil

Procedure: Invite a courier or other delivery person to class. Before the visit, have students write at least one question to ask the speaker.

These questions can be used to initiate a classroom discussion.
- Students ask their questions to encourage the courier to talk about a typical route that he or she follows on an average day.
- Using a local map, students calculate the total number of miles the courier covers.

Act It Out
Materials: U.S. maps, pencils

Procedure: Working in pairs, have students role-play the job of a courier who hand-delivers important packages all over the country.
- One partner is the courier, while the other is the courier's supervisor, or dispatcher.

(Continued on p. 152)

Chapter 12 • 151

(Continued from p. 151)

- The supervisor gives the courier three places for delivery, and together they plan the route and find the actual distance. They decide on the best mode of transportation considering time and cost.
- Then have students exchange roles.

Practice

Have students complete On-The-Job Math 157 *Courier* from the Classroom Resource Binder.

Internet Connection

The following Web sites will give students more information about the job of a courier:

Bureau of Labor Statistics,
Occupational Outlook Handbook
  http://stats.bls.gov/oco/ocos136.htm
Courier Magazine
  http://www.couriermagazine.com

12.6 Problem Solving: Using Proportions
Student Edition pages 270–271

Prerequisite Skills
- Solving proportions
- Solving word problems

Lesson Objectives
- Solve word problems involving proportions.
- Life Skill: Draw a picture to help solve a problem.

Cooperative Group Activity

Write That Problem!

Materials: recipes, scale drawings, maps, supermarket ads, index cards, pencils

Procedure: Have students work in small groups. Give each group one of the following: a recipe, scale drawing, map, or supermarket ad. Different groups should be given different materials.

- Each group writes three word problems involving proportions using the materials provided.
- Groups exchange problems to solve.
- The group that originates the problems checks the solutions. Students may rewrite their problems if others find them difficult to understand or to solve.

Customizing the Activity for Individual Needs

ESL Help students set up proportions with the ratios in the same order. Have them identify the units for each ratio. Then have them write the two ratios showing the units without any numbers. Last, have them fill in the numbers.

Learning Styles Students can:

 draw a picture of the units or objects given in the problem to help solve a problem.

 role-play a situation given in a problem.

 read problems aloud and explain the plan for solving them to a partner.

Reteaching Activity

Have students draw a picture to solve a proportion. Help them to match the parts of the proportion to the pictures. For example, for the proportion $\frac{2}{3} = \frac{5}{?}$, they can begin by drawing 2 squares for every 3 circles to represent the first ratio.

Students can then draw 5 squares and 7.5 circles to represent the second ratio. Help students see that for 1 square you need 1.5 circles.

Alternative Assessment

Students can explain how to solve a problem using proportions.

Example: What steps would you use to solve this problem:

Each day Sheila feeds her 3 kittens 2 cups of cat food. How much cat food would she need each day to feed 4 kittens?

Sample answer: Set up a proportion

Find the cross product and divide by the remaining number.

$4 \times 2 = 8$; $8 \div 3 = 2\frac{2}{3}$ cups

Closing the Chapter
Student Edition pages 272–273

Chapter Vocabulary
Review with students the Words to Know on page 259 of the Student Edition. Then have students quiz each other in pairs.

Have students copy and complete the Vocabulary Review questions on page 272 of the Student Edition.

For more vocabulary practice, have them complete Vocabulary 151 *Ratios and Proportions* from the Classroom Resource Binder.

Test Tips
In small groups, students can explain how to solve a problem by using one of the test tips.

Learning Objectives
Have students review CM6 *Goals and Self-Check* from the Classroom Resource Binder. They can check off the goal they have reached. Note that each section of the quiz corresponds to a Learning Objective.

Group Activity
Summary: Students use multiple unit prices from supermarket ads to find the cost of five items and the number of items they could buy for $10.

Materials: supermarket ads from newspapers, pencils, paper

Procedure: Students can list items that are sold using multiple unit pricing. Then choose the item they would like to buy.

Assessment: Use the Group Activity Rubric on page *xii* of this guide. Fill in the rubric with the additional information below. For this activity, students should have:
- used a proportion to find the cost of five items.
- used a proportion to find the number of items that can be purchased for $10.

RELATED MATERIALS See the unit overview page for other Globe Fearon books that can be used to enrich and extend the material in this unit.

Assessing the Chapter

Traditional Assessment
Chapter Quiz
The Chapter Quiz on pages 272–273 of the Student Edition can be used as either an open-book or closed-book test, or as homework. The quiz can be used to identify concepts in the chapter that students need to review and practice.

Chapter Tests
Use Exercise 159 *Chapter Test A* and Exercise 160 *Chapter Test B Ratios and Proportions* from the Classroom Resource Binder to further assess mastery of chapter concepts.

Additional Resources
Use the Resource Planner on page 146 to assign additional exercises from the Classroom Resource Binder and Workbook.

Alternative Assessment
Real-Life Connection
Display a recipe. Be sure that it indicates the number of servings it makes. Have students do the following:
- Write a ratio to compare two different ingredients.
- Write a ratio to compare the amount of one ingredient with the number of servings.
- Write a proportion to increase the number of servings.
- Solve the proportion to find the amount of the ingredient needed for the increased number of servings.

Interview
Display a map. Ask:
- Why are maps drawn to scale?
- What is the scale of this map?
- What is the scale of this map as a ratio?

Unit Assessment
This is the last chapter in Unit 3, Other Types of Numbers. To assess cumulative knowledge and provide standardized-test practice, administer the practice test on page 274 of the Student Edition and the Unit 3 Cumulative Test, pages T11–12, in the Classroom Resource Binder. These tests are in multiple-choice format. A Scantron sheet is provided on page T2 of the Classroom Resource Binder.

Unit Overview

Unit 4 ▶ Measurement and Geometry

CHAPTER 13 Graphs and Statistics	PORTFOLIO PROJECT Survey Graph	MATH IN YOUR LIFE Making a Budget	USING YOUR CALCULATOR Showing Probability as a Decimal and a Percent	PROBLEM SOLVING Choosing a Scale
CHAPTER 14 Customary Measurement	PORTFOLIO PROJECT Million	ON-THE-JOB MATH Picture Framer	USING YOUR CALCULATOR How Old Are You?	PROBLEM SOLVING Working with Units of Measure
CHAPTER 15 Metric Measurement	PORTFOLIO PROJECT Metric Search	MATH IN YOUR LIFE Better Buy	USING YOUR CALCULATOR Changing Measurements	PROBLEM SOLVING Two-Part Problems
CHAPTER 16 Geometry	PORTFOLIO PROJECT Box	ON-THE-JOB MATH Physical Therapist	USING YOUR CALCULATOR Finding Perimeter and Area	PROBLEM SOLVING Subtracting to Find Area

RELATED MATERIALS

These are some of the Globe Fearon books that can be used to enrich and extend the material in this unit.

Practice & Remediation ▶

Passage to Basic Math
Supply your students with the reinforcement practice they need to develop basic math proficiency.

Test Preparation ▶

Math for Proficiency Level B
Give your students the support they need for success on proficiency exams through this test-taking text.

Real-World Math ▶

Consumer Math
Bring core math skills to life in your class with this rich, consistent program

Planning the Chapter

Chapter 13 • Graphs and Statistics

Chapter at a Glance

SE page	Lesson	
276		Chapter Opener and Project
278	13.1	Pictographs
280	13.2	Single Bar Graphs
282	13.3	Double Bar Graphs
284	13.4	Single Line Graphs
286	13.5	Double Line Graphs
288	13.6	Problem Solving: Choosing a Scale
290	13.7	Circle Graphs
291		Math in Your Life: Making a Budget
292	13.8	Mean (Average)
294	13.9	Median and Mode
296	13.10	Histograms
298	13.11	Probability
299		Using Your Calculator: Showing Probability as a Decimal and a Percent
300		Chapter Review and Group Activity

Learning Objectives

- Read and make pictographs, bar graphs, and line graphs.
- Read circle graphs and histograms.
- Find the mean, median, and mode of a set of data.
- Find simple probability.
- Solve problems about choosing a scale.
- Apply circle graphs to budgets.

Life Skills

- Conduct an opinion survey.
- Make a monthly budget of expenses.
- Adjust a budget to include new expenses.
- Use a calculator to find probability.

Communication Skills

- Choose a scale on a graph to convey a message.
- Use the mean to describe a set of data.
- Write survey questions.
- Write a letter.
- Interview people about important issues.
- Interpret pictographs, bar graphs, line graphs, circle graphs, and histograms.

PREREQUISITE SKILLS

To assess mastery of prerequisite skills, use a selection of exercises from any of the program resources referenced below. The same resources can be used to provide remediation if necessary.

	Program Resources for Review		
Skills	Student Edition	Workbook	Classroom Resource Binder
Rounding whole numbers	Lesson 1.9	Exercise 9	Review 4 Mixed Practice 8
Changing percents to decimals	Lesson 11.2	Exercise 100	
Finding the percent	Lesson 11.9	Exercise 107	Review 142 Mixed Practice 146
Finding the whole	Lesson 11.11	Exercise 109	Review 144 Mixed Practice 146
Finding the part	Lesson 11.3	Exercise 101	Review 138 Mixed Practice 146

Program Resources for Review

Skills	Student Edition	Workbook	Classroom Resource Binder
Adding numbers in columns	Lesson 2.3	Exercise 12	Mixed Practice 18
Dividing larger numbers	Lesson 5.4	Exercise 38	Review 51 Mixed Practice 59
Reducing fractions to lowest terms	Lesson 7.3	Exercise 56	Review 78

Diagnostic and Placement Guide

Chapter 13 Graphs and Statistics

The Percent Accuracy scores are based on the number of problems in a lesson that have been answered correctly.

Resource Planner

After diagnosing your students' needs, use the correlating Program Resources for reinforcement, reteaching, or enrichment. Additional activities for customizing lessons can be found in this guide.

KEY
Reteaching = ⌒   Reinforcement = ↓   Enrichment = ⌒

Lessons		Percent Accuracy				Program Resources			
		50%	65%	80%	100%	↓ Student Edition ▪ Extra Practice	↓ Workbook ▪ Exercises	Teacher's Planning Guide	Classroom Resource Binder
13.1	Pictographs	1	1.3	1.6	2	p. 439	118	⌒ p. 158	↓ Practice 164 *Pictographs*
			(partial credit given)						
13.2	Single Bar Graphs	1	1.3	1.6	2	p. 439	119	↓ p. 159	Visual 11 *Tenths Model* Visual 16 *Blank Table* ↓ Practice 165 *Single Bar Graphs*
			(partial credit given)						
13.3	Double Bar Graphs	1	1.3	1.6	2	p. 439	120	⌒ p. 159	Visual 11 *Tenths Model* ↓ Practice 166 *Double Bar Graphs*
			(partial credit given)						
13.4	Single Line Graphs	1	1.3	1.6	2	p. 440	121	↓ p. 160	↓ Practice 167 *Single Line Graphs*
			(partial credit given)						
13.5	Double Line Graphs	1	1.3	1.6	2	p. 440	122	⌒ p. 161	↓ Practice 168 *Double Line Graphs*
			(partial credit given)						
13.6	Problem Solving: Choosing a Scale	1	1.3	1.6	2	p. 440	123	⌒ p. 162	
			(partial credit given)						
13.7	Circle Graphs	2	2.6	3.2	4	p. 441	124	⌒ p. 163	Visual 12 *Hundredths Model* ↓ Practice 169 *Circle Graphs*
			(partial credit given)						
	Math in Your Life: Making a Budget	Can be used for portfolio assessment.						⌒ p. 163	⌒ Math in Your Life 171 *Making a Budget*
13.8	Mean (Average)	6	8	10	12	p. 441	125	↓ p. 164	↓ Review 162 *Mean (Average)*

		Percent Accuracy			Program Resources			
Lessons	50%	65%	80%	100%	Student Edition Extra Practice	Workbook Exercises	Teacher's Planning Guide	Classroom Resource Binder
13.9 Median and Mode	6	8	10	12	p. 441	126	p. 165	Review 163 *Median and Mode*; Challenge 172 *Using Graphs*
13.10 Histograms	4	5	6	8	p. 442	127	p. 165	Practice 170 *Histograms*
13.11 Probability	2	3	4	5	p. 442	128	p. 166	
Chapter 13 Review								
Vocabulary Review	5	6.5	8	10 (writing is worth 4 points)			p. 168	Vocabulary 161
Chapter Quiz	6	7	9	11			p. 169	Chapter Test A, B 173, 174

Chapter 13

Customizing the Chapter

Opening the Chapter
Student Edition pages 276–277

Photo Activity
Procedure: Have students discuss recycling. Ask about the materials that are recycled in the community. Materials that are commonly recycled are newspapers, magazines, glass, plastic, and tires. Discuss how graphs could be used to illustrate recycling. Ask: *Which kinds of graphs could you use to compare different recycled materials? Which kind of graph could you use to show all the different ways that tires are recycled?*

Answers: bar graph and line graph; circle graph

Words to Know
Review with the students the Words to Know on page 277 of the Student Edition. You can help students remember these terms by having them group related terms together under a general heading, such as "Graph-Related" and "Data-Related."

The following words and definitions are covered in this chapter.

data information gathered from surveys or experiments

graph a visual display that shows data in different ways; includes bar, line, and circle graphs

pictograph a graph that uses pictures to represent data

mean sum of the data divided by the number of data; also called *average*

median middle number when data are ordered from least to greatest

mode number or numbers that appear most often in a set of data

histogram a graph that shows how many times an event occurred

probability the chance that an event will occur

Survey Graph Project
Summary: Students create graphs based on data gathered from a class survey.

Materials: paper, pencils, construction paper, colored pencils or markers, tape

Procedure: Distribute materials to students. Students can complete this project in class (over one to three days). Or, students can write their survey questions and five possible choices as homework and post their surveys for classmates to complete. Then students can use the results to create their graphs as homework.

Assessment: Use the Individual Activity Rubric on page *xi* of this guide. Fill in the rubric with the additional information below. For this project, students should have:

- collected data for a survey question.
- made a graph of the data collected.

Learning Objectives

Review with the students the Learning Objectives on page 277 of the Student Edition before starting the chapter. Students can use the list of objectives as a learning guide. Suggest that they write the objectives in a journal or use CM6 *Goals and Self-Check* from the Classroom Resource Binder.

After each lesson, have students write an example of the skill they learned under the appropriate objective. Suggest that students use these notes as a learning guide to help them study for the chapter test.

13.1 Pictographs
Student Edition pages 278–279

Prerequisite Skills

- Rounding whole numbers
- Comparing whole numbers
- Multiplying whole numbers by 10, 100, 1,000

Lesson Objectives

- Make pictographs.
- Communication Skill: Read and interpret pictographs.

Words to Know

data, graph, pictograph

Cooperative Group Activity

Tracking Attendance

Materials: local newspapers, large pieces of poster board, markers

Procedure: Gather local newspapers that span several weeks. Organize students into small groups. Give each group several newspapers, a large piece of poster board, and markers.

- Each group chooses a local sports team from the newspaper. The group looks through the newspapers to find the attendance for the last five games.
- Group members work together to create a pictograph showing the attendance for the last five games. The pictograph must include a key.
- Display the pictographs in the classroom. Have a volunteer from each group explain its graph to the class.

Customizing the Activity for Individual Needs

ESL To increase students' understanding of the word *graph*, have students list words that have *graph* in them, such as *graphics*, *pictograph*, and *photograph*. Point out that each word refers to a visual display that gives information.

Learning Styles Students can:

 write the number represented by each picture in a pictograph next to the picture. They then add the numbers to find the total number for each category.

 place counters on construction paper to make a pictograph.

 say the name of each category out loud. Then ring a buzzer or bell once for each picture in a pictograph.

Enrichment Activity

Have students draw a pictograph of the data in the following table. Suggest that they use ☺ to represent 5 days and discuss how they might represent 8 days.

Sunny Days	
Month	Number of Days
January	15
February	10
March	25
April	8

Alternative Assessment

Students can use counters to make a pictograph.

Example: There are 250 girls and 400 boys at summer camp. Use counters to make a pictograph of this data. How many campers should each counter represent? How many counters represent the number of girls? the number of boys?

Possible answers: If each counter represents 25 campers, 10 counters represent the number of girls, 16 counters represent the number of boys. If each counter represents 10 campers, 25 counters represent the number of girls, 40 counters represent the number of boys.

13.2 Single Bar Graphs
Student Edition pages 280–281

Prerequisite Skills
- Finding multiples of whole numbers
- Comparing whole numbers

Lesson Objectives
- Make single bar graphs.
- Communication Skill: Read and interpret single bar graphs.

Cooperative Group Activity

How Long Is Your Trip?

Materials: Visual 16 *Blank Table,* Visual 11 *Tenths Model* or lined paper, pencils, colored pencils or markers

Procedure: Organize students into groups of equal numbers. Give each group one copy of Visuals 16 and 11.

- Group members record in a table the number of minutes it takes each student in the group to get to school.
- They then make a bar graph of the data in the table. Each bar should represent the time it takes a student to get to school. The graph can be horizontal or vertical.
- Be sure that students include a title for the graph and labels for the axes. Students can decorate the borders of the graph.
- Display the bar graphs in the classroom. Ask a volunteer from each group to explain the graph to the class.

Customizing the Activity for Individual Needs

ESL Describe the term *horizontal* as *across*; describe the term *vertical* as *up and down*. Have students point to horizontal and vertical objects in the room.

Learning Styles Students can:

 use graph paper to make bar graphs. Remind them that each square is equal to the same number of units.

 cut out strips of grid paper to model bars. They then can paste the strips to create a bar graph.

 skip count the numbers on the scale aloud as they draw the bars.

Reinforcement Activity

Have students make a data table from the information in the graph below.

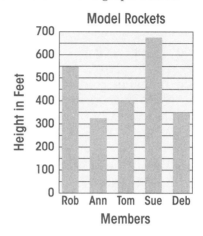

Alternative Assessment

Students can answer questions about making a single bar graph.

Bake Sale	
Goods Sold	Amount Collected
Cookies	$225
Pies	$75
Cakes	$150
Donuts	$200

Example: Look at the data in the table above. What type of scale will you use to show the numbers? Suppose that each interval on the scale is $25. How many intervals do you need to show $75? $225?

Answers: Accept all reasonable responses; 3 intervals; 9 intervals

13.3 Double Bar Graphs
Student Edition pages 282–283

Prerequisite Skills
- Finding multiples of whole numbers
- Comparing whole numbers

Lesson Objectives
- Make double bar graphs.
- Communication Skill: Read and interpret double bar graphs.

Cooperative Group Activity

Oh That Population!

Materials: Visual 16 *Blank Table,* Visual 11 *Tenths Model* or lined paper, pencils, colored pencils or markers, index cards, a world almanac or a resource for state population figures

Procedure: Organize the class into groups of four students. Give each group one copy of both Visuals 16 and 11. Give each student a pencil and an index card. Provide access to a world almanac.

- Each group member chooses a state and finds the population for that state in 1900 and in 1950. The student then writes a short paragraph about that state on an index card.
- Group members compile the information on the population of the states in a three-column table. They then make a double bar graph.
- Be sure that students include a title for the graph, labels for the axes, and a key. Students can decorate the borders of the graph with images that represent each state. Include the index cards about the states in the display.

Customizing the Activity for Individual Needs

ESL Be sure that students understand how to use the *key* in a double bar graph. Have students match the colors in the key to the color on each bar.

Learning Styles Students can:

 color code the columns in the data table to match the colors in the key and bars.

 use a straight edge to line up numbers along the axis and the bars when reading a graph.

 have a partner read him or her the data from the bar graph to check the numbers against the data in the table.

Enrichment Activity

Provide students with a piece of lined paper. Have them use the data in the following table to draw a double bar graph.

Hours Worked in March		
Name	Paper Route	Mowing Lawns
Don	11	14
Sandy	8	21
Tomás	6.5	16

Discuss the scale for the graph. Ask, *What scale would you use? Why?*

Possible answer: If 1 unit represents 1 hour, you could use $6\frac{1}{2}$ units to represent $6\frac{1}{2}$ hours.

Alternative Assessment

Students can interpret a double bar graph by writing a question about the graph such as the one pictured below from the Student Edition.

Example: Look at the double bar graph in the Everyday Problem Solving section on page 283 of your book. Write a new question about the graph. Write the answer to your question.

Possible answer: For which movie was the least number of tickets sold? Movie 1 at 5:00.

13.4 Single Line Graphs
Student Edition pages 284–285

Prerequisite Skill
- Comparing whole numbers

Lesson Objectives
- Make single line graphs.
- Communication Skill: Read and interpret single line graphs.

Cooperative Group Activity

Our Armed Forces

Materials: almanacs, poster board, paper, markers, self-stick notes, Visual 12 *Hundredths Model* or graph paper, Visual 16 *Blank Table*

Procedure: Locate the pages in four almanacs that show the number of personnel on active duty in the U.S. Army, Navy, Marines, and Air Force. Place a self-stick note on those pages in each almanac. Organize students into four groups. Give an almanac, sheet of poster board, Visuals 12 and 16 or paper, and markers to each group. Assign each group only one division.

- Each group turns to the corresponding marked page of the almanac. Group members make a table showing the total number of enlisted personnel in their assigned division for each year for the past five years.
- Group members then make a line graph using the data in the table.
- A volunteer from each group explains its graph to the class.

Customizing the Activity for Individual Needs

ESL To help students differentiate between the types of graphs, have them draw an example of a line, a bar, and two bars together. Under each, have the student say and then write the name of the graph that would use the picture to visually display data.

Learning Styles Students can:

 use two different colors to highlight increases and decreases on line graphs.

 use string and pushpins in a corkboard to model line graphs.

 listen to a partner read the data in the line graph aloud from left to right. For example: "In January, there was 4.5 inches of rainfall."

Reinforcement Activity

Ask students to make a line graph using the following data about Jeannie's baby:

Age	Weight (kg)
Birth	3
3 months	6.5
6 months	7.5
9 months	8.5
1 year	10

Alternative Assessment

Students can show their understanding of line graphs by summarizing the information in a line graph.

Example: Find a line graph in a newspaper and write a paragraph summarizing the information in the graph.

Possible answer: This graph shows U.S. unemployment rates over the past five years.... Unemployment has decreased each year since 1995.

13.5 Double Line Graphs
Student Edition pages 286–287

Prerequisite Skill
- Comparing whole numbers

Lesson Objectives
- Make double line graphs.
- Communication Skill: Read and interpret double line graphs.

Cooperative Group Activity

Comparing Temperatures

Materials: almanacs, poster board, paper, markers, self-stick notes, Visual 12 *Hundredths Model*, Visual 16 *Blank Table*

Procedure: Locate the pages in each almanac that show the normal monthly temperatures and precipitation for different cities. Place a self-stick note on those pages. Organize students into groups of five or six. Give an almanac, sheet of poster board, Visuals 12 and 16 or paper, and markers to each group.

- Each group turns to the marked pages of the almanac. They choose two cities and find the normal monthly temperatures for each city for the same six months.
- Group members compile the information in a three-column table. They then make a double line graph using the data from the table.
- Students discuss how the graph shows change over time and how the graph compares the weather conditions in the two cities.
- Ask what a traveler might wear in each city.

Customizing the Activity for Individual Needs

ESL Help students interpret double line graphs. Have them pose questions that can be answered by reading the graph. Then have them answer their own questions.

Learning Styles Students can:

 shade each column in the data table with the same color they use for drawing the lines representing the data.

 trace each line graph with a finger when reading and interpreting the graphs.

 explain aloud how double line graphs show change over time and describe what data are being compared.

Enrichment Activity

Ask students to add the normal monthly temperatures for a third city to the line graph they made in the Cooperative Group Activity. Have them write a report about the information in the graphs.

Alternative Assessment

Students can demonstrate their understanding of double line graphs by describing the steps they take to make a double line graph.

Example: Write out the steps you used to make the graph in Practice Exercise 1 on page 287 in the Student Edition.

Answer: Accept all reasonable answers. Use page 286 in the Student Edition as a reference.

13.6 Problem Solving: Choosing a Scale *Student Edition pages 288–289*

Prerequisite Skills
- Reading and making bar graphs
- Reading and making line graphs

Lesson Objectives
- Communication Skill: Choose a scale on a graph to convey a message.

Cooperative Group Activity

Compare the Graphs

Materials: newspapers, magazines, paper, pencils, Visual 12 *Hundredths Model* or graph paper

Procedure: Have students find examples of bar graphs or line graphs from newspapers or magazines. Organize students into groups of three or four. Assign each group one of the graphs with which to work. Give each group two sheets of graph paper or two copies of Visual 12.
- Each group draws a graph of the same data. They use a scale with intervals that are greater than those shown on the original graph.
- Each group draws another graph of the same data. They use a scale with intervals that are less than those shown on the original graph.
- Volunteers from each group present their graphs and the original graph to the class. They describe how the intervals on the scale affect the appearance of the graph for the same set of data.

Customizing the Activity for Individual Needs

ESL To help students understand that the two graphs show the same data, have them write the data from each graph each in a different table. Then have them compare the data in the tables to see that they are the same. Also have them explain the connection between the scale and its purpose.

Learning Styles Students can:

 copy the graphs on graph paper to help understand the differences between the scales of the graphs. Have each square in both graphs equal the same number of units.

 use stacking cubes or tiles to represent and compare different bar graphs.

 describe to a partner the differences in the scales of the two graphs. The partner then repeats what was said so the first student can determine if his or her differences are correct.

Enrichment Activity

Have students write questions for the double line graph below, which is from the Problem Solving Strategy section on page 289 of the Student Edition. Then have students exchange their questions and solve.

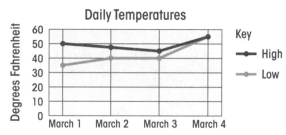

Alternative Assessment

Students can show their understanding of choosing a scale by describing how a change in scale affects the message the graph presents.

Example: Make a data table with two entries. Draw two bar graphs. Make one graph with intervals that are one unit apart and the other with intervals that are more than one unit apart. Does the change between bars seem greater when the intervals are one unit apart or more than one unit apart?

Answer: Accept all reasonable graphs. Students should note that the change seems greater when the intervals are 1 unit apart.

13.7 Circle Graphs
Student Edition page 290

Prerequisite Skills
- Finding the whole
- Finding the part
- Changing percents to decimals

Lesson Objective
- Read circle graphs.

Cooperative Group Activity

Circle Graph "Talk"
Materials: newspapers, magazines, large paper plates, pencils, markers, tape

Procedure: Organize students into groups of five or six. Give newspapers, magazines, pencils, markers, tape, and a paper plate to each group.
- Each group finds a circle graph in one of the sources. They copy the graph onto a paper plate.
- In turn, each group tapes the circle graph to the chalkboard. Group members take turns describing different sections of the graph.
- One group member summarizes the information in the graph for the class.
- Students organize the circle graphs into logical categories of subject matter and group them accordingly on a bulletin board labeled *Circle Graph Talk*.

Customizing the Activity for Individual Needs
ESL Remind students that *percent* is a part based on 100. Have them add up the percents in each circle graph to see that the total is 100%. Have them find the total of the parts to see if it adds up to the whole number the graph represents.

Learning Styles Students can:

 find and write down the numbers associated with each part of a circle graph. Then add the parts to check that they make up the whole.

 cut apart a circle graph and rearrange the sections in order from greatest to least or least to greatest.

 explain orally what each section of a graph represents.

Enrichment Activity
Provide students with copies of each of the graphs below which show family budgets.

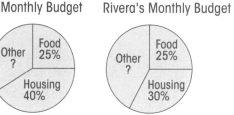

Ask, *Does each family spend the same amount on food? Why or why not? Does each family spend the same amount on housing? Why or why not? What percent of the monthly income does each family spend on other?*

Answers: No; the Smith family spends $450 and the Rivera family spends $600; yes; each family spends $720 on housing; 35% for the Smith family, 45% for the Rivera family.

Alternative Assessment
Students can demonstrate their understanding of circle graphs by answering questions about the graph below.

Example: What percent does each part of this circle graph show? If the whole graph represents 120 students, how many students does the shaded part represent?

Answers: 50%; 60 students

MATH IN YOUR LIFE
Making a Budget
Student Edition page 291

Lesson Objectives
- Apply circle graphs to budgets.
- Life Skill: Make a monthly budget of expenses.
- Communication Skill: Write a letter to a school, city, town, county, or state treasurer.
- Life Skill: Adjust a budget to include a new expense.

(Continued on p. 164)

(Continued from p. 163)

Activities

Tools of the Trade
Materials: small record books, pencils

Procedure: Have students keep a record of their weekly income and expenses. They write down:
- how much money they begin the week with either from allowance or earnings.
- what they spend their money on and how much they spend.

At the end of the week, they:
- put their expenses into a few major categories.
- calculate the total of each category.
- divide the total of each expense category by their total income to determine the percent of the whole that each represents.

Students then consider how they could use this information to help plan a budget for the next week. Ask, *How would you adjust your budget if you know there will be additional expenses?*

Write a Business Letter
Materials: paper and pens or computer word processor, computer graphics

Procedure: Students write letters asking for information about how the yearly school, city, town, county, or state budget is calculated. They find out what the total expected income is and the estimated expenses. Students then:
- calculate the percent that each expense is of the total expected income.
- use a computer to generate a circle graph.
- write a report about the budget.

Act It Out
Materials: credit card offers

Procedure: Students discuss the results of spending more than a person earns. Discuss:
- the use of credit cards.
- different credit card rates.
- ways to manage credit.

Practice

Have students complete Exercise 171 *Making a Budget* in the Classroom Resource Binder.

Internet Connection

The following Web sites have features that can be used as extension activities:
  The National Budget Simulation
    http://garnet.berkeley.edu:3333/budget/budget.html.
  The Wedding Garden
    http://www.weddingarden.com

13.8 Mean (Average)
Student Edition pages 292–293

Prerequisite Skills
- Adding numbers in columns
- Dividing larger numbers

Lesson Objectives
- Find the mean of a set of data.
- Communication Skill: Use the mean to describe a set of data.

Word to Know
mean

Cooperative Group Activity

Average Highs and Lows
Materials: local newspapers spanning two weeks, paper, pencils

Procedure: Organize students into small groups. Have newspapers available in a central location.
- Each group chooses a sports team or player. Groups use the newspapers to find scores for a particular team or player in its chosen sport.
- Working together, group members use the data to find the average number of runs, goals, or points for the team or player over the span of two weeks.
- Each group presents its findings to the class.

Customizing the Activity for Individual Needs

ESL To help students understand the term *mean*, relate it to the word *average*. The *average* student grade is a C, because it is in the middle of all the grades. Note that the word *mean* has many different meanings in common English usage. Help students to distinguish between the mathematical use of the word and its common uses.

Learning Styles Students can:

 highlight key words in each step when finding the mean.

 use counters to represent each number. Then find the mean by combining counters and dividing them into equal groups based on the number counted.

 say the steps out loud when finding the mean.

Reinforcement Activity

Give students local newspapers from the past four days. Have them find the average high temperature and the average low temperature for that period.

Alternative Assessment

Students can find the mean by using counters.

Example: Show the numbers 2, 1, 8, 6, and 3 with different sets of counters. Use the counters to find the mean of 2, 1, 8, 6, and 3.

Answer: Students combine the counters and divide them into five equal groups to find a mean of 4. Alternatively, students stack the counters and move them from place to place until all stacks are the same height with four in each stack.

13.9 Median and Mode
Student Edition pages 294–295

Prerequisite Skills
- Ordering whole numbers
- Comparing whole numbers

Lesson Objectives
- Find the median and mode of a set of data.
- Find the mean, median, and mode from a graph.

Words to Know
median, mode

Cooperative Group Activity

Comparing Heights

Materials: yardsticks, rulers, or tape measures with inches, paper, pencils

Procedure: Organize the class into two groups. Give a yardstick, ruler, or tape measure to each group.
- Students measure and record the height of each person in their group to the nearest inch.
- Group members work together to find the median height and the mode height of their group. They then present their findings to the class.

Customizing the Activity for Individual Needs

ESL Help students understand the terms *median* and *mode*. Have students make bars of different lengths using locking cubes. Have them order the bars from shortest to longest. Then identify the *middle* bar as the *median*. Have students roll a number cube 10 times, writing down the number rolled each time. Ask which number came up most often. Identify that number as the *mode*.

Learning Styles Students can:

 write numbers in order from least to greatest vertically down the page instead of from left to right across the page when finding the median.

 write the numbers on self-stick notes and organize the notes from least to greatest to find the median. Group the notes with the same number to find the mode.

 read the numbers out loud to make sure they are organized from least to greatest when finding the median. Read the numbers aloud when finding the mode.

Reinforcement Activity

Provide each student with 10 index cards. Have them write a number on each card, mix them up, and exchange cards with a partner. Have each student find the median and mode for the set of numbers on the index cards.

Alternative Assessment

Students can find the median for an even number of data and recognize that data can have more than one mode.

Example: Find the median and mode of the following set of numbers: 5, 6, 2, 3, 7, 6, 2, 8.

Answer: Median is 5.5; modes are 2 and 6.

13.10 Histograms
Student Edition pages 296–297

Prerequisite Skill
- Reading and making bar graphs

Lesson Objective
- Read histograms.

Word to Know
histogram

Cooperative Group Activity

Histogram Exchange

Materials: Visual 11 *Tenths Model* or graph paper, construction paper, markers

Procedure: Organize students into an equal number of small groups. Provide each group with a copy of Visual 11, construction paper, and markers. Place the following table on the chalkboard.

Weekly Lunch Costs	
Amount Spent	Number of Students
$0 – $9	25
$10–$19	60
$20–$29	85
$30–$39	45
$40–$49	10

(Note that the amount spent for lunch each week is divided into $10 intervals. You might wish to point out that a student who spends $4 will be in the first interval; that the 25 students for this interval may spend anywhere from $0 to $9 a week for lunch.)

- Have each group make a histogram for the data in the table and paste it on a sheet of construction paper. They can decorate the sides of the histogram with pictures of nutritious lunch items.
- Each group thinks of four questions to ask about the histogram. They write them on the construction paper below the histogram.
- Groups exchange histograms and answer each other's questions about the histogram.

Customizing the Activity for Individual Needs

ESL To help students understand the term *histogram*, have one example of each type of graph as visuals. Have students explain what each bar or dot represents on all graphs, using keywords such as *between, over, below, most frequent*. When the students get to the histogram, emphasize that all the bars are touching and that each bar represents the number of items that occur between two numbers. The other graphs have bars and dots that represent the number of items for one number or object.

Learning Styles Students can:

 label the bars of the histogram with the corresponding numbers.

 use tiles to represent the bars in a histogram.

 say the number that corresponds to each bar out loud.

Enrichment Activity

Challenge students to create their own histograms. Have them survey the class to find the number of hours their classmates spend watching TV each night (0–1, 1–2, 2–3, 3–4, etc.). Have students make a histogram to display their results.

Alternative Assessment

Students can demonstrate their understanding of histograms by writing questions about a histogram such as the one pictured below from the Student Edition.

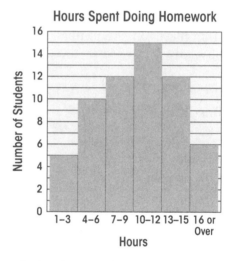

Example: Look at the histogram shown. Write three new questions with answers about the histogram using the terms *between, less than,* and *more than.*

Answer: Accept all reasonable questions. Be sure that answers are correct to questions.

13.11 Probability
Student Edition page 298

Prerequisite Skills
- Writing ratios
- Reducing fractions to lowest terms

Lesson Objective
- Find simple probability.
- Life Skill: Use a calculator to find probability.

Word to Know
probability

166 • Chapter 13

Cooperative Group Activity

Probability Race

Materials: pencils, eight index cards

Procedure: Organize students into two teams. Distribute four index cards to each team.

- Each student in the class prints his or her name on the chalkboard.
- Students imagine that each name is on a card in a box and he or she can pick a name from the box without looking.
- Have the students answer the following question: *What is the probability that you will pick a name that begins with the letter B?*
- Team members write their own four probability questions about the situation each on a separate index card.
- Have a student collect the cards and mix them up. Then have another student pick a card and read the question to the class.
- Team members work together to find the answer. As soon as a team agrees upon the answer, they should raise their hands.
- The first team to answer the question correctly scores one point.
- Continue asking questions until one team wins by scoring 5 points.

Customizing the Activity for Individual Needs

ESL Explain that *outcomes* are all possible results. Have students list the outcomes that could occur by rolling a number cube. Give them examples of what might be considered *favorable outcomes*, such as "rolling a 3; or rolling an even number." Have students say the number of favorable outcomes, using the word *outcomes* in their statement.

Learning Styles Students can:

 highlight the favorable outcomes.

 roll number cubes to explore probability.

 say the numbers or choices to partners, who then repeat the choices back to the student.

Enrichment Activity

Have students write the letters of their first names on separate slips of paper. Ask them to determine the probability of picking a particular letter of their choice. Then have a student place the letters in a box and pick one without looking. Have him or her record the result and replace the letter. Ask the student to repeat this process 10 or more times. Ask the student to calculate how often he or she picked the chosen letter out of all the letters picked. This is to see how close the experimental probability is compared to the theoretical probability determined earlier.

Alternative Assessment

Students can show their understanding of probability by solving probability riddles.

Example: I have a spinner with 5 blue sections, 2 purple sections, and 3 yellow sections. The probability of landing on my color is $\frac{1}{2}$. What color am I?

Answer: blue

USING YOUR CALCULATOR
Showing Probability as a Decimal and a Percent
Student Edition page 299

Lesson Objectives

- Find simple probability.
- Life Skill: Use a calculator to find probability as a decimal and as a percent.

Activities

Spinners Game

Materials: Visual 17 *Spinners* from the Classroom Resource Binder or commercial spinner, calculators, pencils, paper

Procedure: Have students work in pairs. Give each pair of students a different spinner. Students can write a number in each section of their spinner or color each section. They may repeat numbers or colors.

- One student writes a probability question about the spinner.
- The other student writes the answer to the question as a fraction and then uses the calculator to find the probability as a decimal and as a percent.
- Students then reverse roles.

Variation: One student can give a possible probability for the spinner as a fraction, decimal, or percent. The other student writes a question whose answer is the given probability.

Roll That Cube!

Materials: number cube or six index cards, calculators, pencils, paper

(Continued on p. 168)

(Continued from p. 167)

Procedure: Have students work in pairs. Give each pair a number cube or six index cards. Have students number the cube or the cards. They may repeat numbers.
- One student writes a probability question about tossing the number cube or picking a number card without looking.
- The other student writes the answer to the question as a fraction. The student then uses the calculator to find the probability as a decimal and as a percent.
- Students then reverse roles.
- Students may compare the theoretical probability they found with an experimental probability for one of the questions: for example, *the probability of tossing a 3 on the number cube.* They can toss the number cube 18 times and record the results. Have them count the number of times a 3 comes up. They divide that number by 18. Have them compare the quotient to the theoretical probability they found for this event.

Closing the Chapter
Student Edition pages 300–301

Chapter Vocabulary

Review with the students the Words to Know on page 277 of the Student Edition. Then have students quiz each other in pairs.

Have students copy and complete the Vocabulary Review questions on page 300 of the Student Edition.

For more vocabulary practice, have them complete Vocabulary 161 *Graphs and Statistics* from the Classroom Resource Binder.

Test Tips

Have pairs of students take turns showing how to solve a problem by using one of the test tips.

Learning Objectives

Have the students review CM6 *Goals and Self-Check* from the Classroom Resource Binder. They can check off the goal they have reached. Note that each section of the quiz corresponds to a Learning Objective.

Group Activity

Summary: Students conduct a survey on a school issue. They organize the data and find the mean, median, and mode. Then create a graph of the data for display.

Materials: pencils, paper, poster board, markers

Assessment: Use the Group Activity Rubric on page *xii* of this guide. Fill in the rubric with the additional information below. For this project, students should have:
- calculated the mean, median, and mode correctly.
- made an appropriate graph for the data.

RELATED MATERIALS See the unit overview page for other Globe Fearon books that can be used to enrich and extend the material in this unit.

Assessing the Chapter

Traditional Assessment

Chapter Quiz
The Chapter Quiz on pages 300–301 of the Student Edition can be used as either an open- or closed-book test, or as homework. The quiz can be used to identify concepts in the chapter that students need to review and practice.

Chapter Tests
Use Chapter Test A Exercise 173 and Chapter Test B Exercise 174 *Graphs and Statistics* from the Classroom Resource Binder to further assess mastery of chapter concepts.

Additional Resources
Use the Resource Planner on page 155 of this guide to assign additional exercises from the Classroom Resource Binder and Workbook.

Alternative Assessment

Follow Instructions
Write the following table of data on the chalkboard:

Paintings Sold	
Year	Number Sold
1995	80
1996	85
1997	95
1998	90
1999	105

Ask students to follow these instructions:
- Draw a pictograph of the data.
- Draw a bar graph of the data.
- Find the mean, median, and mode for the data set.
- Find the number of paintings sold between 1995 through 1999.

Answers: Check students' graphs; mean 91; median 90; no mode; 455 paintings

Real-Life Connection
Distribute newspapers and magazines containing graphs to the students. Ask students to pick one graph to present to the class. Have them:
- tell what the graph is about.
- identify the scale of the graph or the total amount if it is a circle graph.
- state an important fact the graph illustrates.
- compare the data in the graph.

Planning the Chapter

Chapter 14 • Customary Measurement

Chapter at a Glance

SE page	Lesson
302	*Chapter Opener and Project*
304	14.1 Length
306	14.2 Weight
308	14.3 Capacity (Liquid Measure)
310	14.4 Time
311	*Using Your Calculator: How Old Are You?*
312	14.5 Problem Solving: Working with Units of Measure
314	14.6 Elapsed Time
316	14.7 Temperature
317	*On-The-Job Math: Picture Framer*
318	*Chapter Review and Group Activity*

Learning Objectives
- Change units of length, weight, capacity, and time.
- Find elapsed time.
- Find changes in temperature.
- Solve problems using customary measurements.
- Apply measurement to framing.

Life Skills
- Choose the best unit for measuring everyday objects.
- Use a calculator to find a person's age.
- Use a calculator to convert units.
- Draw pictures to help solve problems.
- Calculate elapsed time in everyday situations.

Communication Skills
- Explain how to convert units of measurement.
- Explain how to find elapsed time.
- Read a thermometer.
- Describe temperature changes.

Workplace Skills
- Calculate the amount of wood needed for a picture frame.
- Determine the cost of framing a picture.

PREREQUISITE SKILLS
To assess mastery of prerequisite skills, use a selection of exercises from any of the program resources referenced below. The same resources can be used to provide remediation if necessary.

Skills	Student Edition	Workbook	Classroom Resource Binder
Multiplying and dividing whole numbers	Lessons 4.2, 5.2	Exercises 27, 36	
Multiplying mixed numbers	Lesson 8.4	Exercise 68	Review 89 Practice 94 Mixed Practice 96
Multiplying fractions and whole numbers	Lessons 8.1, 8.2	Exercises 65–67	Review 88 Practice 94 Mixed Practice 96
Adding and subtracting whole numbers	Lessons 2.2, 3.2	Exercises 11, 19	

Program Resources for Review

Diagnostic and Placement Guide

Chapter 14 Customary Measurement

The Percent Accuracy scores are based on the number of problems in a lesson that have been answered correctly.

Resource Planner

After diagnosing your students' needs, use the correlating Program Resources for reinforcement, reteaching, or enrichment. Additional activities for customizing lessons can be found in this guide.

KEY
Reteaching = ⤴   Reinforcement = ⬇   Enrichment = ⤴

Lessons	Percent Accuracy				Program Resources			
	50%	65%	80%	100%	Student Edition • Extra Practice	Workbook • Exercises	Teacher's Planning Guide	Classroom Resource Binder
14.1 Length	9	12	14	18	p. 443	129	⬇ p. 172	Visual 16 *Blank Table* ⬇ Review 176 *Length*
14.2 Weight	10	13	16	20	p. 443	130	⤴ p. 173	⬇ Review 177 *Weight*
14.3 Capacity (Liquid Measure)	9	12	14	18	p. 444	131	⤴ p. 174	Visual 16 *Blank Table* ⬇ Review 178 *Capacity (Liquid Measure)* ⤴ Challenge 184 *Estimating Measurement*
14.4 Time	4	5	6	8	p. 444	132	⬇ p. 175	Visual 16 *Blank Table* ⬇ Review 179 *Time*
14.5 Problem Solving: Working with Units of Measure	1	1	2	2		133	⤴ p. 176	⬇ Practice 181 *Problem Solving: Working with Units of Measure*
14.6 Elapsed Time	6	8	10	12	p. 445	134	⤴ p. 176	⬇ Review 180 *Elapsed Time*
14.7 Temperature	2	3	3	4	p. 445	135	⬇ p. 177	⬇ Practice 182 *Temperature*
On-The-Job Math: Picture Framer	*Can be used for portfolio assessment.*							⤴ On-The-Job Math 183 *Picture Framer*
Chapter 14 Review								
Vocabulary Review	5	6.5	8	10	(writing is worth 4 points)		⬇ p. 178	⬇ Vocabulary 175
Chapter Quiz	10	13	16	20			⬇ p. 179	⬇ Chapter Test A, B 185, 186

Customizing the Chapter

Opening the Chapter
Student Edition pages 302–303

Photo Activity

Procedure: Discuss measuring length using customary units of measure. The customary units used to measure length are the inch, foot, yard, and mile. Ask, "Which unit would you use to measure a book?" "A room?" "The distance between two cities?" Discuss why it would be important to measure weight. Introduce the customary units ounce, pound, and ton. Ask, "Which unit is used to measure the weight of a car?" "An apple?" Finally, discuss the customary units used to measure capacity: fluid ounce, pint, quart, and gallon. Ask students to identify items that are usually in quart containers.

Words to Know

Review with the students the Words to Know on page 303 of the Student Edition. You can help students remember these words by suggesting that they write words they cannot yet define on index cards. They then write the definitions on the reverse sides.

The following words and definitions are covered in this chapter.

length how long an object is

Customary System of Measurement measurement units used in the United States

weight how heavy an object is

capacity how much space is in a container

elapsed time the amount of time that has passed between two given times

degrees (°) units used to measure temperature

Million Project

Summary: Students calculate the number of days for a million seconds, the volume (in gallons) of a million drops of water, and the weight of a million pennies.

Materials: measuring cups, eyedropper, pennies, scale, water, paper, pencils

Procedure: Provide students with measuring tools and calculators. Students can work on this project at home or in school. Help students get started by suggesting that they find the number of drops of water in 1 cup, the number of seconds in a day, or the weight of 100 pennies.

Encourage oral discussion to help students develop a plan by thinking through the steps. Allow students to use calculators. Have students make posters to display their results.

Fill in the rubric with the additional information below. For this project students should have:

- found that it would take about $11\frac{1}{2}$ days to equal a million seconds.
- found that a million drops of water is about 250 gallons of water.
- found the weight of a million pennies is about 5,000 pounds.

Learning Objectives

Review with the students the Learning Objectives on page 303 of the Student Edition before starting the chapter. Students can use the list of objectives as a learning guide. Suggest that they write the objectives in a journal or use CM6 *Goals and Self-Check* in the Classroom Resource Binder.

After each lesson, have students write an example of the skill they learned under the appropriate objective. Suggest that students use these notes as a learning guide to help them study for the chapter test.

14.1 Length
Student Edition pages 304–305

Prerequisite Skills
- Multiplying whole numbers
- Multiplying mixed numbers
- Dividing by whole numbers

Lesson Objectives
- Change customary units of length.
- Life Skill: Choose the best unit of length for measuring a given object.

Words to Know
length, Customary System of Measurement

Cooperative Group Activity

How Long Is It?

Materials: 12-inch rulers, yardsticks, measuring tapes, paper, pencils, Visual 16 *Blank Table*

Procedure: Organize students into groups of three. Make a list on the board of 10 objects around the room, school, and schoolyard. The objects should have a wide range of lengths, from 1 inch to several yards. Give each student two copies of Visual 16.

- Each student in the group makes a three-column chart with headings *Object, Estimated Length,* and *Actual Length.* Have them copy the list of objects from the chalkboard in the first column in their charts and discuss the appropriate unit to be used to measure each object.
- Each group member estimates the length of each object and records the estimate on his or her chart.
- Next, group members measure the length of each object and record the measurement on their tables. The member whose estimate is closest to the actual measurement scores a point.
- The student in each group with the most points after all 10 objects have been measured is the winner.
- To extend the activity, have students categorize the measured objects by length. Write the

following headings on the chalkboard: *less than 1 foot, between 1 foot and 3 feet, greater than 1 yard.* Volunteers from each group write the name of an object under the appropriate heading.

Customizing the Activity for Individual Needs

ESL To help students understand the word *yard* as a unit of measure, demonstrate with a 12-inch ruler and a yardstick.

12 inches = 1 foot
36 inches = 3 feet
3 feet = 1 yard

Learning Styles Students can:

 highlight the unit of measure and make a list of objects that are measured in inches, yards, and miles.

 cut pieces of string that are 1 inch, 1 foot, and 1 yard in length.

 orally explain how to recognize what unit to use to measure an object.

Enrichment Activity

Discuss reasons why someone would need to change from one unit of measure to another. Have students find the total amount paid for the following order of wood trim.

Wood Trim Order		
Length	Cost	Amount Paid
12 inches	$6 per foot	
18 inches	$6 per foot	
24 inches	$5 per foot	
30 inches	$5 per foot	
36 inches	$7 per yard	
	Total	

Answers: $6; $9; $10; $12.50; $7; $44.50

Alternative Assessment

Students can explain how to change units of length.

Example: A room is 12 feet long. Explain how you can change this measurement to (a) yards and (b) inches.

Answers: (a) divide 12 by 3 to change feet to yards; (b) multiply 12 by 12 to change feet to inches.

14.2 Weight
Student Edition pages 306–307

Prerequisite Skills
- Multiplying whole numbers
- Multiplying mixed numbers
- Dividing whole numbers

Lesson Objectives
- Change customary units of weight.
- Life Skill: Choose the best unit of weight for measuring a given object.

Word to Know
weight

Cooperative Group Activity

Ordering by Weight

Materials: several sets of 5 classroom objects of different weights, such as a textbook, a notebook, a pencil, an eraser, and chalk; scale; paper; pencils

Procedure: Organize students into small groups. Provide a set of 5 objects and a scale for each group.

- Group members feel the weight of each object. They then order the 5 objects by weight from lightest to heaviest. They write the name of each object in that order.
- Group members weigh each object. They write the name of the object and its corresponding weight on paper. Then they order the 5 objects from lightest to heaviest and compare this order with their estimate.
- Groups compare their lists.

Customizing the Activity for Individual Needs

ESL Explain the difference between *pound* as a verb, as in *to pound the desk*, and *pound* as a noun, *a pound of cheese*.

Learning Styles Students can:

 make flash cards of equivalent unit measures to use for changing between units of weight.

 write measurements on index cards and arrange the cards in order from least to greatest. Cards with equivalent measurements can be placed one on top of the other.

 listen to a tape-recorded explanation of how to change from one unit of weight to another.

Chapter 14 • 173

Reteaching Activity

Help students become familiar with the customary units of weight by letting them hold objects that weigh close to 1 ounce, 1 pound, and 5 pounds.

Alternative Assessment

Students can explain how to change units of weight.

Example: Vincent's grandfather weighs 200 pounds. Explain how you can change his weight into (a) ounces and (b) tons.

Answers: (a) Multiply 200 by 16 to change pounds to ounces; (b) divide 200 by 2000 to change pounds to tons.

14.3 Capacity
Student Edition pages 308–309

Prerequisite Skills
- Multiplying whole numbers
- Multiplying mixed numbers
- Dividing whole numbers

Lesson Objectives
- Change customary units of capacity.
- Life Skill: Choose the best unit of liquid measure for measuring a given capacity.

Word to Know
capacity

Cooperative Group Activity

Capacity Spin
Materials: Visual 16 *Blank Table*, Visual 17 *Spinners*, paper, pencils

Procedure: Have students work in pairs. Give each pair a copy of Visuals 16 and 17. Draw the following on the chalkboard.

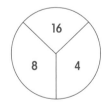 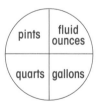

- Each pair of students makes two spinners as shown above and a chart with the following column heads: *fluid ounces, pints, quarts, gallons*.
- One student spins both spinners. The result is a measurement. For example, 16 fluid ounces.
- The other student writes the measurement on the chart in the appropriate column.
- Students work together to fill in the rest of the row in their chart by changing the measurement into the other three units of measure—1 pint, $\frac{1}{2}$ quart, $\frac{1}{8}$ gallon.
- Students take turns spinning the spinners. They work together to change the measurements into different units of measure. (There are 12 different possible measurements.)
- Pairs of students can exchange their charts and check each other's work.

Customizing the Activity for Individual Needs

ESL To help students learn the meaning of the words *pint, quart,* and *gallon,* bring in a 1-pint container, a 1-quart container, and a 1-gallon container. Students fill each with water as they say each word.

Learning Styles Students can:

 draw diagrams to represent relationships among the units, so it is easier to decide whether to multiply or divide.

 use clean empty containers of each unit as a reference.

 orally ask themselves "Am I looking for a larger number or a smaller number?" when changing from one unit to another.

Reinforcement Activity

Have students bring in clean, empty containers that have labeled measurements. Allow students to present a demonstration to the class explaining how to change the larger unit to the smaller unit of capacity. One student can show the calculations on the chalkboard. Another student can fill the larger container with sand and then pour the sand into the smaller container(s) to verify the results.

Alternative Assessment

Students can identify the correct operation to change units of liquid measure.

Example: A water jug holds 3 quarts. What operation do you use to change this capacity into (a) fluid ounces, (b) pints, and (c) gallons? Explain.

Answers: (a) multiply 3 by 32 since ounces are smaller than quarts; (b) multiply 3 by 2 since pints are smaller than quarts; (c) divide 3 by 4 since gallons are larger than quarts.

14.4 Time
Student Edition page 310

Prerequisite Skills
- Multiplying whole numbers, fractions, and mixed numbers
- Dividing whole numbers

Lesson Objective
- Change from one unit of time to another unit.

Cooperative Group Activity

Recipe Rewards
Materials: cookbooks, paper, pencils, Visual 16 Blank Table

Procedure: Organize students into pairs. Give each pair a cookbook and a copy of Visual 16.

- Students use the cookbook to look up the roasting or baking times per pound for turkey, chicken, beef, and pork.
- Pairs calculate the time needed to cook a 20-pound turkey, a 6-pound chicken, a 7-pound beef roast, and a 9-pound pork roast.
- Pairs write their answers on a chart. The chart should show the number of hours and minutes needed to roast each type of meat.

Customizing the Activity for Individual Needs
ESL Have students make a chart of words by writing the English word for a unit of time and the word in their native language for that same unit.

Learning Styles Students can:

 use a highlighter to match equivalent measures of time.

 use a stopwatch to get a sense of how long a given number of minutes or seconds is.

 explain aloud how to convert from one unit of measure to another.

Reinforcement Activity

Have students bring in a CD or cassette tape and write down the titles and lengths of the songs (usually given as minutes:seconds). Have them convert each time into seconds, then find the total length (in seconds) of all the songs. Have them change this total back to minutes to find the total length of the CD or cassette.

Alternative Assessment
Students can find the number of minutes in a given time using the classroom clock.

Example: How many minutes are in $1\frac{1}{2}$ hours?

Answer: 90 minutes

USING YOUR CALCULATOR
How Old Are You?
Student Edition page 311

Lesson Objectives
- Life Skill: Use a calculator to determine a person's age in various units of time.
- Life Skill: Use a calculator to convert units of time.

Activities

Animal Longevity
Materials: almanac or reference book on animals, poster board, art pencils, calculators

Procedure: Organize students into small groups. Provide materials for each group.

- Using an almanac or other reference book, students find the longevity of their favorite animal. They then use calculators to find the expected life span in days and in minutes.
- Students create an Animal Longevity bulletin board display. They mount or draw pictures of the chosen animals on poster board and include in their display the animals' longevity in years, days, and minutes.

Variation: Students create a bulletin-board display of the ages (in days) of family members.

Space Flights
Materials: reference books on space flights, calculators

Procedure: Organize students into small groups. Provide materials for each group.

- Students use reference books to find the duration of several space flights. They then calculate to find the length of each flight in years, days, hours, or minutes.
- Students write their results on the chalkboard for comparison.

14.5 Problem Solving: Working with Units of Measure
Student Edition pages 312–313

Prerequisite Skills
- Multiplying whole numbers, fractions, and mixed numbers
- Dividing whole numbers

Lesson Objectives
- Solve word problems with customary units of measure.
- Life Skill: Draw a picture to help solve a problem.

Cooperative Group Activity

Food Problems
Materials: empty food boxes (cereal, rice, sugar); cans; beverage bottles; paper; pencils

Procedure: Organize students into groups of three. Give each group several food containers: either boxes from *dry goods* or cans and bottles from *liquid goods*.
- Students make a list of the items that could be in their containers. Next to each item they write the measure on the container: either the weight or the capacity.
- Group members work together to create three word problems about the items on their list. For example, *What is the total weight of 5 boxes of cereal?* At least one problem must require a change of units. Students write each problem on one side of a sheet of paper. They write the solutions on the backs.
- Groups exchange and solve each other's problems. They check their work by looking at the solutions.

Customizing the Activity for Individual Needs
ESL Have students prepare self-help cards with each of the conversion facts from Lessons 14.1–14.5. These may be used as flash cards.

Learning Styles Students can:

 highlight different units in different colors to help decide which units to regroup.

 use real-life objects to model problems.

 read word problems aloud to identify different units of measure used. Explain whether answers make sense.

Enrichment Activity
Use the items from the Cooperative Group Activity and include the prices of the items. Have students make up word problems involving the cost of items.

Alternative Assessment
Students can identify and name the appropriate fact needed to change one unit of measure to another when solving word problems.

Example: Ellen has 3 quarts of milk. Each serving is 6 fluid ounces. What fact does she need to find the number of servings in the 3 quarts of milk?

Answer: 1 quart = 32 fluid ounces

14.6 Elapsed Time
Student Edition pages 314–315

Prerequisite Skill
- Subtracting whole numbers

Lesson Objectives
- Determine elapsed time.
- Life Skill: Calculate elapsed time in real-life situations.

Word to Know
elapsed time

Cooperative Group Activity

Travel Times
Materials: plane, train, or bus schedules; paper; pencils

Procedure: Have students work in pairs. Give each pair a schedule from an airline, bus, or train.
- Students use the schedule to determine the departure and arrival times from your city to or from five other cities.
- Students use these times to calculate the length of each trip one-way.

Customizing the Activity for Individual Needs
ESL Many countries use military time (24-hour time) for all scheduling. Explain that A.M. is the abbreviation for *ante meridiem*, meaning *before noon*; P.M. is the abbreviation for *post meridiem*, meaning *after noon*. Have students list the times including A.M. and P.M., at which certain events occur during the day.

Learning Styles Students can:

 draw clock faces to find the elapsed time.

 move hands on an actual clock to determine elapsed time.

 count up from the time given to find the answer.

Enrichment Activity

Have students research time zones around the world. Have them determine a specific time across different time zones. Students can present an oral report of their findings.

Alternative Assessment

Students can use the demonstration clock to help them find elapsed time.

Example: A movie started at 7:35 P.M. and ended at 9:50 P.M. How long was the movie? Use a clock to help you.

Answer: 2 hours 15 minutes

14.7 Temperature
Student Edition page 316

Prerequisite Skills
- Adding whole numbers
- Subtracting whole numbers

Lesson Objectives
- Calculate changes in temperature.
- Communication Skill: Read a thermometer.
- Communication Skill: Describe temperature changes.

Cooperative Group Activity

Hot and Cold

Materials: thermometers, plastic cups, clock, paper, pencils, hot water, cold water

Procedure: Organize students into groups of three or four. Give each group a thermometer and two plastic cups.

- Students label one cup "warm" and the other cup "cold."
- Students fill the "warm" cup with warm tap water. They measure the temperature of the water and record it on paper.
- Students fill the other cup with cold tap water. They measure and record the temperature.
- Students predict what each temperature will be in 30 minutes. They decide if the temperatures will be the same or whether the temperature of the water in one cup will change more than the other.
- After 30 minutes, students measure the temperature of the water in each cup again. They record their results.
- Students calculate the change in temperature for each cup over the 30-minute period. They discuss the results.

Customizing the Activity for Individual Needs

ESL Help students understand the Fahrenheit thermometer. Compare the Fahrenheit scale to the Celsius scale used in many countries.

Learning Styles Students can:

 use two different colors to model temperature changes.

 use a ribbon to model temperature change on a thermometer model.

 explain when to add and when to subtract to find temperature change.

Reinforcement Activity

Have students choose a city in the United States they would like to visit. Have them use reference books to find and record the city's average high and low temperatures. Then have them calculate the change in temperature.

Alternative Assessment

Students can find a change in temperature using models of two thermometers.

Example: Use models to find the change in temperature from 25° to 73°.

Answer: 48°

ON-THE-JOB-MATH
Picture Framer
Student Edition page 317

Lesson Objectives
- Apply measurement to framing.
- Communication Skill: Interview a picture framer.
- Workplace Skill: Determine the amount of wood needed for a picture frame.
- Workplace Skill: Determine the cost of framing a picture.

(Continued on p. 178)

(Continued from p. 177)

Activities

Look It Up
Materials: none
Procedure: Have students visit an art-supply or picture-framing store. Have them find out:
- the size of standard frames and their costs.
- the cost per foot for having a frame custom-made.

Have them use this information to compare the costs of buying a pre-made frame and having a frame made to order.

Conduct an Interview
Materials: none
Procedure: Have students interview a picture framer at a local art-supply or picture-framing store. Ask them to find out:
- what types of wood are used in picture framing.
- the costs of different types of woods.
- the dimensions of the largest and smallest frames.
- how the framer became interested in picture framing.
- what math is used on the job.

Act It Out
Materials: art reference books, art-supply catalogs
Procedure: Organize students into small groups. Provide each group with an art reference book and an art-supply catalog.
- Students use the reference books to find the dimensions of works of art created by famous artists: such as Picasso, Van Gogh, or others.
- Students find a frame from the art-supply catalog that would be suitable for framing the masterpiece or they calculate the cost of making a frame for painting.

Practice

Have students complete the On-The-Job Math Exercise 183 *Picture Framer* in the Classroom Resource Binder.

Internet Connection

Students can find information about frame shops and framing, as well as professional associations, schools, and trade journals related to the job of picture framer, at
The Art & Framing Headquarters
http://www.artframing.com/index.htm

Closing the Chapter
Student Edition pages 318–319

Chapter Vocabulary

Review with the students the Words to Know on page 303 of the Student Edition. Then have students quiz each other in pairs.

Have students copy and complete the Vocabulary Review questions on page 318 of the Student Edition.

For more vocabulary practice, have them complete Vocabulary 175 *Customary Measurement* from the Classroom Resource Binder.

Test Tips

Have pairs of students take turns reading one of the test tips and showing how to solve a problem by using that tip.

Learning Objectives

Have students review CM6 *Goals and Self-Check* from the Classroom Resource Binder. They can check off the goal they have reached. Note that each section of the quiz corresponds to a Learning Objective.

Group Activity

Summary: Students plan a picnic and calculate the total weight or capacity of the picnic food.

Materials: supermarket ads from newspapers, paper, pencils

Assessment: Use the Group Activity Rubric on page *xii* of this guide. Fill in the rubric with the additional information below. For this project, students should have:
- decided on foods to serve and amounts of each item needed.
- listed all items and amounts with their appropriate weights or capacities, and find the total.

RELATED MATERIALS See the unit overview page for other Globe Fearon books that can be used to enrich and extend the material in this unit.

Assessing the Chapter

Traditional Assessment

Chapter Quiz
The Chapter Quiz on pages 318–319 of the Student Edition can be used as an open-book test, closed-book test, or as homework. The quiz can be used to identify concepts in the chapter that students need to review and practice.

Chapter Tests
Use Chapter Test A Exercise 185 and Chapter Test B Exercise 186 *Customary Measurement* from the Classroom Resource Binder to further assess mastery of chapter concepts.

Additional Resources
Use the Resource Planner on page 171 of this guide to assign additional exercises from the program.

Alternative Assessment

Interview
Display a grocery item showing the weight or capacity in a smaller unit than usually found on the label. Ask:
- Is the unit of measure on the label the best unit for this item? Why or why not?
- What is the total weight or capacity of 3 of these items?

As-Large-As-Life Numbers
Provide students with rulers, scales, and thermometers. Have students do the following:
- Measure the length and width of a desk in feet.
- Change the dimensions to inches.
- Record the temperature of the classroom.
- Find the temperature with a rise of 7°.
- Find the weight of a book in pounds.
- Change the weight to ounces.

Planning the Chapter

Chapter 15 • Metric Measurement

Chapter at a Glance

SE page	Lesson
320	Chapter Opener and Project
322	15.1 What Is the Metric System?
323	Math in Your Life: Better Buy
324	15.2 Length
326	15.3 Mass
328	15.4 Capacity (Liquid Measure)
330	15.5 Comparing Metric and Customary Measurements
331	Using Your Calculator: Changing Measurements
332	15.6 Problem Solving: Two-Part Problems
334	Chapter Review and Group Activity

Learning Objectives

- Identify prefixes used in the metric system.
- Change metric units of length, mass, and liquid capacity.
- Compare metric and customary units of measurement.
- Solve problems using metric measurements.
- Apply measurement to finding the better buy.

Life Skills

- Compare prices to find the better buy.
- Measure length in centimeters and meters.
- Compare metric capacities.
- Use a calculator to change customary measures to metric measures.
- Make a table.

Communication Skills

- Describe customary and metric measures found in ads and on food labels.
- Explain how unit price labeling can help a person choose the better buy.
- Read nutrition fact labels.
- Write a guide on using the metric system.

PREREQUISITE SKILLS

To assess mastery of prerequisite skills, use a selection of exercises from any of the program resources referenced below. The same resources can be used to provide remediation if necessary.

	Program Resources for Review		
Skills	Student Edition	Workbook	Classroom Resource Binder
Understanding place value to thousands	Lesson 1.3	Exercise 3	Review 2 Mixed Practice 8
Reading and writing decimals	Lesson 10.2	Exercise 85	Review 120
Multiplying whole numbers by 10, 100, 1,000	Lesson 4.7	Exercise 32	
Dividing larger numbers	Lesson 5.4	Exercise 38	Review 51

Skills	Program Resources for Review		
	Student Edition	Workbook	Classroom Resource Binder
Multiplying decimals by 10, 100, 1,000	Lesson 10.8	Exercise 91	
Dividing decimals by 10, 100, 1,000	Lesson 10.11	Exercise 94	
Solving proportions	Lesson 12.3	Exercise 114	Review 153

Diagnostic and Placement Guide

**Chapter 15
Metric Measurement**

The Percent Accuracy scores are based on the number of problems in a lesson that have been answered correctly.

Resource Planner

After diagnosing your students' needs, use the correlating Program Resources for reinforcement, reteaching, or enrichment. Additional activities for customizing lessons can be found in this guide.

KEY
Reteaching = ↶   Reinforcement = ↓   Enrichment = ↷

Lessons	Percent Accuracy				Program Resources			
	50%	65%	80%	100%	↓ Student Edition Extra Practice	↓ Workbook Exercises	Teacher's Planning Guide	Classroom Resource Binder
15.1 What Is the Metric System?	2	3	3	4		136	↓ p. 182	
Math in Your Life: Better Buy	Can be used for portfolio assessment.						↶ p. 183	↶ Exercise 193
15.2 Length	6	8	10	12	p. 446	137	↶ p. 184	↓ Review 188 *Length*
15.3 Mass	6	8	10	12	p. 446	138	↓ p. 185	↓ Review 189 *Mass*
15.4 Capacity (Liquid Measure)	6	8	10	12	p. 446	139	↶ p. 185	↓ Review 190 *Capacity (Liquid Measure)* ↶ Challenge 194 *Estimating Measurement*
15.5 Comparing Metric and Customary Measurements	4	5	6	8		140	↶ p. 186	
15.6 Problem Solving: Two-Part Problems	1	1	2	2		141	↶ p. 187	Visual 22 *Problem-Solving Steps* ↓ Review 191 *Problem Solving: Two-Part Problems* ↓ Practice 192 *Problem Solving: Two-Part Problems*
Chapter 15 Review								
Vocabulary Review	4.5	5.5	7.5	9			↓ p. 189	↓ Vocabulary 187
	(writing is worth 4 points)							
Chapter Quiz	10	13	16	20			↓ p. 189	↓ Chapter Test A, B 195, 196

Customizing the Chapter

Opening the Chapter
Student Edition pages 320–321

Photo Activity
Have students look at the photo of the speed limit sign on the George Washington Bridge. This sign gives the speed limit in both metric and customary units, 45 mph and 70 km/hr. Ask a volunteer to read the photo caption. Before students answer the question, ask them what *mph* and *km/hr* stand for. Draw a large rectangle on the chalkboard and write the following text in it: Rochester: 5 mi 8 km. Tell students to suppose that this is a road sign on the highway. Ask them to identify which measurement is a metric measurement and which is a customary measurement. Then have students brainstorm and think of other times when they use or see customary and metric measurements.

Words to Know
Review with the students the Words to Know on page 321 of the Student Edition. Help students remember these words by suggesting that students write a sentence using each of them.

The following words and definitions are covered in this chapter.

metric system the system of measurement based on the number 10 that is used in most countries

meter the basic unit used to measure length

gram the basic unit used to measure weight

liter the basic unit used to measure liquid capacity

unit price the cost of one item or one unit measure of an item

Metric Search Project
Summary: Students find examples of metric measurements and how they are used. They explain how metric and customary measurements compare.

Materials: newspapers, magazines, food labels, paper, pencils

Procedure: Have students find magazines, newspapers, and food labels that contain customary and metric measurements. Students may cut out and paste the advertisements or food labels on a poster or in a journal. Ask them to organize the information by type, length, weight or mass, and capacity. Encourage students to discuss why some items may be measured in customary or metric units, or both. Have students compare the items to develop a sense of size in each system. Students may complete this project in small groups in class or as homework.

Assessment: Use the Individual Scoring Rubric on page xi of this guide. Fill in the rubric with the additional information below. For this project, students should have:
- found at least five varied types of ads or labels.
- correctly organized the information according to the directions.

Learning Objectives
Review with the students the Learning Objectives on page 321 of the Student Edition before starting the chapter. Students can use the list of objectives as a learning guide. Suggest that they write the objectives in a journal or use CM6 *Goals and Self-Check* from the Classroom Resource Binder.

After each lesson, have students write an example of the skill they learned under the appropriate objective. Suggest that students use these notes as a learning guide to help them study for the chapter test.

15.1 What Is the Metric System?
Student Edition page 322

Prerequisite Skills
- Understanding place value to thousands
- Reading and writing decimals

Lesson Objective
- Identify prefixes used in the metric system.

Words to Know
metric system, meter, gram, liter

Cooperative Group Activity

Metric Measure Race
Materials: paper, pencils, index cards

Procedure: The day before, organize students into small groups. Have students prepare questions and answers like the Practice questions on page 322 of the Student Edition: for example, "How many grams in a hectogram?" "How much of a meter is a millimeter?" Students may write the questions on one side of an index card with the answers on the other side. Have groups exchange questions and check each other's work, then hand in the cards.

182 • Chapter 15

On the day of the activity, write the following prefixes and values at the top half of the chalkboard for reference:

kilo	hecto	deka	deci	centi	milli
1,000	100	10	.1	.01	.001

Then write **Team 1** and **Team 2** on the side of the chalkboard. Organize the students into teams. Pick an index card and read the prepared questions.

- The students on each team may consult among themselves. Then a volunteer from one of the teams writes the answer to the question on the chalkboard. Teams receive a point for each correct answer.
- The game continues until all students have had a turn at the chalkboard. The team with the highest score wins.

Customizing the Activity for Individual Needs

ESL To increase students' understanding of the words *meter*, *gram*, and *liter*, show students a meter stick, a paper clip, and a liter bottle of soda. Use these items to discuss *length*, *weight*, and *capacity*.

Learning Styles Students can:

 In one color, highlight the prefixes greater than 1 unit (*kilo*, *hecto*, *deka*) and in another color highlight the prefixes less than 1 unit (*deci*, *centi*, *milli*).

 create flash cards of metric prefixes and their values.

 take turns saying the prefixes and their corresponding values aloud with a partner.

Reinforcement Activity

Distribute flash cards of metric prefixes and their values (the front shows the prefix, the back shows the value) to students. Have student pairs divide the cards in half and take turns identifying the value of each prefix.

Alternative Assessment

Students can solve riddles containing values to identify prefixes in the metric system.

Example: One of me is the same as 1,000 meters. What am I?

Answer: a kilometer

MATH IN YOUR LIFE
Better Buy
Student Edition page 323

Lesson Objectives
- Apply measurement to finding the better buy.
- Life Skill: Find the better buy for items measured in metric units.
- Communication Skill: Explain the advantages of unit price labeling.
- Communication Skill: Write a letter to a person in food services about finding the best buys.

Words to Know
unit price

Activities

Take a Class Trip
Materials: grocery store flyer, paper, pencils
Procedure: Organize the class in small groups. Have students plan to shop for an award celebration for 50 people. They will need to purchase healthy snacks, juice, and paper products. Give each group a store flyer or take the class to a local grocery store. Have the members of each group:
- prepare a shopping list of the items they need and the quantity of each item.
- look at the unit pricing labels (usually found on the grocery shelves) for different sizes of the same items on their lists. Determine which sizes are the best buys for each item and explain why.
- decide whether unit pricing is the only factor to consider. For example, are some sizes more convenient to use than others?
- write down the cost of all the items they decide to purchase, including their measurements.
- calculate the total cost and the cost per person for the award celebration.

Write a Business Letter
Materials: pens, writing paper, envelopes, stamps
Procedure: Have students write letters to the head of food services at your school or the owners of local restaurants. In their letters have them find out:
- how the person decides what to purchase.
- whether the person orders larger quantities of some items and smaller quantities of others, and why.
- how the person finds the best buys for the cafeteria or restaurant.

(Continued on p. 184)

(Continued from p. 183)

Act It Out
Materials: none
Procedure: Have student pairs role-play the following situation in which one student is a grocery store clerk and the other is a customer who needs brown sugar. The customer notes that he or she can buy a 400-gram box for $.59 or a 700-gram bag for $.79. The customer is not sure which to buy and asks the clerk for advice.
- Students consider what kinds of questions the clerk could ask the customer before making a recommendation.
- Students make a recommendation and explain why they gave that advice.

Practice
Have students complete the Math in Your Life Exercise 193 *Better Buy* from the Classroom Resource Binder.

Internet Connection
The following Web sites can be used to shop for groceries and other products online:

Groceronline
http://www.groceronline.com/consumer/GOLFrameset.asp

netgrocer
http://www.netgrocer.com

15.2 Length
Student Edition pages 324–325

Prerequisite Skills
- Identifying prefixes used in the metric system
- Multiplying whole numbers and decimals by 10, 100, 1,000
- Dividing whole numbers and decimals by 10, 100, 1,000

Lesson Objectives
- Change metric units of length.
- Life Skill: Measure length in centimeters and meters.

Cooperative Group Activity

Find Your Height
Materials: meter sticks, chalk, pencils, paper
Procedure: Organize students into small groups. Give each group a meter stick and piece of chalk.

- Students take turns standing against the chalkboard to measure each other's heights. The heights are recorded on a chart.
- Students then change the height measurement in centimeters to meters.
- To extend the activity, students find the mean, median, and mode for their group.
- You may wish to have students find the mean for the class and compare this with the mean for each group.

Customizing the Activity for Individual Needs
ESL To increase understanding that 1,000 *millimeters* is the same as 1 *meter*, not larger (because 1,000 is larger than 1). Have them draw one line that is 1,000 millimeters long and one that is 1 meter long. Students compare the lines.

Learning Styles Students can:

write each metric unit of length vertically down a piece of paper in order from greatest to least. They then can draw arrows and indicate the operation needed to change from larger to smaller units or smaller to larger units.

move arrows one unit at a time until they reach the desired unit.

refer to the Useful Facts chart on page 324 of the Student Edition and read facts aloud to help answer questions.

Enrichment Activity
Have students organize the class data collected in the cooperative group activity in a frequency table and then make a histogram. The height could be organized into 25-cm intervals from 0–24 cm to 150–199 cm.

Alternative Assessment
Students can explain how to change metric units of length.

Example: Tell how you can change 2.3 kilometers to meters. How many meters are in 1 kilometer? Would you multiply or divide? Why?

Answer: There are 1,000 meters in 1 kilometer. Multiply 2.3 by 1,000 because you are changing from a larger unit to a smaller unit.

15.3 Mass
Student Edition pages 326–327

Prerequisite Skills
- Identifying prefixes used in the metric system
- Multiplying whole numbers and decimals by 10, 100, 1,000
- Dividing whole numbers and decimals by 10, 100, 1,000

Lesson Objectives
- Change metric units of mass.
- Life Skill: Identify metric measures of mass on nutrition fact labels.

Cooperative Group Activity

Find Your Group

Materials: index cards, pencils

Procedure: Organize students into groups of four. Give each group four index cards.
- Each group creates a set of index cards with equivalent values of mass. For example, one set could be 14 grams, 1.4 dekagrams, 1,400 centigrams, 14,000 milligrams.
- A volunteer collects all the index cards and shuffles them. Another volunteer distributes one card to each student in the class.
- Students find classmates that have cards with the same value as theirs, forming groups of equivalent measurements.

Customizing the Activity for Individual Needs

ESL To help students understand that *1,000 grams* is the same as *1 kilogram*, provide students with metric weights and a pan balance so they can compare the measures.

Learning Styles Students can:

 draw diagrams to represent relationships among the units.

 find objects around the classroom or at home with masses of about 1 gram, 5 grams, and 1 kilogram.

 record the Useful Facts on page 326 of the Student Edition on a tape recorder and replay it as needed.

Reinforcement Activity
Distribute labels from cereal boxes, canned goods, or other packaged foods. For every amount on the label given in grams, have students write the number of milligrams. For every amount on the label given in milligrams, have them write the number of grams.

Alternative Assessment
Students can explain how to change metric units of mass.

Example: Tell how you can change 500 centigrams to grams. How many centigrams are in 1 gram? Would you multiply or divide? Why?

Answer: There are 100 centigrams in 1 gram. Divide 500 by 100 because you are changing from a smaller unit to a larger unit.

15.4 Capacity (Liquid Measure)
Student Edition pages 328–329

Prerequisite Skills
- Identifying prefixes used in the metric system
- Multiplying whole numbers and decimals by 10, 100, 1,000
- Dividing whole numbers and decimals by 10, 100, 1,000

Lesson Objectives
- Change metric units of capacity.
- Life Skill: Compare the capacity of beverage containers.

Cooperative Group Activity

Metric Match Game

Materials: index cards, pencils

Procedure: Give each student six index cards and a marker. Organize students into groups of three.
- Each student makes three pairs of cards: one card shows a metric measurement of capacity; the other shows an equivalent measurement: for example, 1 kiloliter and 1,000 liters.
- One person from each group mixes up the cards and places them facedown in an array.
- Group members take turns picking two cards. If the cards show equivalent measurements, they keep the cards. If the measurements are not equivalent, the student puts them back in their places facedown.

- Students continue until there are no cards left. The student in each group with the most cards at the end wins.

Customizing the Activity for Individual Needs

ESL To help students differentiate between *1 liter* and *1 milliliter*, bring in a one-liter bottle of soda. Compare this to a drop of water, which is about *1 milliliter*.

Learning Styles Students can:

 write metric capacity equivalents on a poster to use as a reference for the classroom.

 use a graduated cylinder or metric measuring cup to determine the capacities of different containers.

 ask themselves "Am I looking for a larger number or a smaller number?" when changing from one unit to another.

Enrichment Activity

Challenge students to answer the following question: How many kiloliters is 1 million milliliters?

Answer: 1 kiloliter

Ask students to explain how they arrive at their answer.

Alternative Assessment

Students can change metric measurements of capacity by finding an amount for every unit.

Example: Find the number of kiloliters, centiliters, and milliliters in 200 liters.

Answers: .2 kiloliter; 20,000 centiliters; 200,000 milliliters

15.5 Comparing Metric and Customary Measurements
Student Edition page 330

Prerequisite Skill
- Recognizing equivalent relationships

Lesson Objective
- Compare metric and customary units of measurement.

Cooperative Group Activity

Comparing Measurements

Materials: two almanacs, yardstick or tape measure, several different-sized containers labeled in quarts or gallons, paper, pencils, poster board for each group

Procedure: Organize students into four groups. Give an almanac to each of two groups, a yardstick to the third, and the containers to the fourth. Write the following relationships on the chalkboard:

    2.5 centimeters is about 1 inch.
    1 liter is about 1 quart.
    1.6 kilometers is about 1 mile.
    28 grams is about 1 ounce.

- The group with one almanac finds the lengths of the world's 5 longest rivers in miles and changes the lengths to kilometers by multiplying the miles by 1.6.
- The group with the yardstick measures each other's heights in inches, then finds the approximate heights in centimeters by multiplying by 2.5.
- The group with the other almanac finds the weights of the world's 5 smallest animals in ounces and changes the weights to grams by multiplying by 28 grams.
- The group with the different-sized containers finds the approximate number of liters each container will hold.
- Each group creates a poster of their findings. A volunteer explains the work to the class.

Customizing the Activity for Individual Needs

ESL Help students understand the term *about* when it is used in a sentence such as "2.5 centimeters is about 1 inch." Explain that *about* can also mean *not exactly*. Have students compare 2.5 centimeters to 1 inch on a ruler.

Learning Styles Students can:

 highlight metric units in one color and customary units in another color.

 use tools that are calibrated in both customary and metric units: for example, a ruler and a measuring cup.

 explain the relationship between metric and customary units of measure to a partner.

186 • Chapter 15

Reteaching Activity

Have students practice measuring the length, mass, and capacity of items using both metric and customary measuring tools. Have them measure an item first using metric measurement, then have them re-measure using customary measurement. Have them record the results.

Alternative Assessment

Students can use estimation to compare and convert units from one system to another.

Examples: If 1 mile is about 1.6 kilometers, 3 miles is about how many kilometers?

If 2 inches is about 5 centimeters, 1 foot is about how many centimeters?

Answers: 3 miles is about 5 kilometers; since 1 foot equals 12 inches, 1 foot is about 30 centimeters.

USING YOUR CALCULATOR
Changing Measurements
Student Edition page 331

Lesson Objectives
- Life Skill: Use a calculator to change customary measures to metric measures.
- Life Skill: Compare customary measures with metric measures.

Activities
Change That Measurement!
Materials: Visual 17 *Spinners*, calculators, paper, pencils

Procedure: Organize students into groups of three players. Display a copy of the following conversion chart, which can be found on page 331 of the Student Edition.

```
inches x 2.54   → centimeters
feet x 30.48    → centimeters
yards x .9144   → meters
miles x 1.609   → kilometers
pints x .4732   → liters
quarts x .9464  → liters
gallons x 3.785 → liters
pounds x .4536  → kilograms
```

Provide each group with a copy of Visual 17. Have members of the group make the following spinners which they will use for the activity.

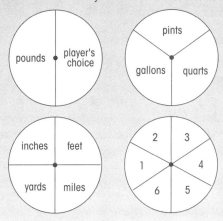

- Each player picks a different spinner with measurement units.
- Players take turns spinning their measurement unit spinner and the number spinner. They combine the results and write a customary measurement on a sheet of paper: for example, 4 quarts.
- Students then change their customary measurement to a metric measurement using the conversion chart and a calculator. Students must record the results: for example, 4 quarts × .9464 → 3.7856 liters.
- Students check each other's work.
- Players then exchange unit spinners so that each will have an opportunity to work with units of length, weight, and capacity.
- If a player spins *player's choice,* the player may choose any one of the other units on the spinners to combine with the number spun.

15.6 Problem Solving: Two-Part Problems
Student Edition pages 332–333

Prerequisite Skill
- Understanding metric units of length, mass, and liquid capacity

Lesson Objectives
- Solve word problems using metric measurements.
- Life Skill: Make a table to help solve a problem.

Cooperative Group Activity

Write Your Own Problem

Materials: paper, pencils; food items

Procedure: Organize students into groups of four. Have students bring in clean boxes or cans from food items that have metric units on the label. Have students write the price of the item on the label.

- Group members place their items in the center of their table.
- One student writes the first part of a word problem that involves one or more of the food items, then passes the problem to the student on the right.
- This student writes the second part of the word problem, then passes the problem to the student on the right.
- The third student solves the problem.
- The fourth student checks the answer.
- The members of the group then reflect on their work. They may change the problem to make it easier or more complex.
- Students reverse roles and repeat the activity.
- All word problems could be collected and stored in a central location. Students may pick a word problem to solve at a later date.
- The food items could be donated to a community food pantry.

Customizing the Activity for Individual Needs

ESL Students may have trouble understanding some of the words in a word problem. Ask them to underline the words they don't know, then use a dictionary to look up the meaning of each word.

Learning Styles Students can:

 draw pictures to represent different parts of the word problem.

 fill in the steps on Visual 22 *Problem-Solving Steps* as they solve metric two-part problems.

 read the word problem into a tape recorder and replay it one part at a time as he or she solves the problem.

Reinforcement Activity

Have students copy and complete the table below to solve the following problem.

Jo's trail mix has 1.5 kilograms of raisins for every 3.5 kilograms of nuts. How many kilograms of raisins and how many kilograms of nuts are needed to make 20 kilograms of trail mix?

Raisins	1.5	?	4.5	?
Nuts	3.5	7	?	?
Trail Mix	5	?	?	?

Answers: 6 kilograms raisins; 14 kilograms nuts

Alternative Assessment

Students can change units of metric measure to solve word problems.

Example: How much longer is a metal strip that measures 1.2 meters than a metal strip that measures 894 millimeters? Tell what measurement needs to be changed. Then solve.

Possible answer: Change 1.2 meters to 1,200 millimeters. Then subtract; 306 millimeters

Closing the Chapter
Student Edition pages 334–335

Chapter Vocabulary
Review with the students the Words to Know on page 321 of the Student Edition. Then have students quiz each other in pairs.

Have students copy and complete Vocabulary Review questions on page 334 of the Student Edition.

For more vocabulary practice, have them complete Vocabulary 187 *Metric Measurement* from the Classroom Resource Binder.

Test Tips
Have pairs of students take turns showing how to solve a problem by using one of the test tips.

Learning Objectives
Have students review CM6 *Goals and Self-Check*. They can check off the goal they have reached. Note that each section of the quiz corresponds to a Learning Objective.

Group Activity
Summary: Students write a guide for tourists to use while traveling that explains how to use the metric system.

Materials: construction paper, paper, pencils, markers or crayons, staplers

Procedure: Students can describe the units of measure and where they will be used. Students may draw or cut out and paste appropriate pictures to illustrate the guide.

Assessment: Use the Group Activity Rubric on page xii of this guide. Fill in the rubric with the additional information below. For this project, students should have:

- described the metric units of length, mass, and capacity.
- described the relationship among the units of measure and given examples.

RELATED MATERIALS See the unit overview page for other Globe Fearon books that can be used to enrich and extend the material in this unit.

Assessing the Chapter

Traditional Assessment
Chapter Quiz
The Chapter Quiz on pages 334–335 of the Student Edition can be used as either an open-book test, closed-book test, or as homework. Use the quiz to identify concepts in the chapter that students need to review and practice.

Chapter Tests
Use Chapter Test A Exercise 195 and Chapter Test B Exercise 196 *Metric Measurement* from the Classroom Resource Binder to further assess mastery of chapter concepts.

Additional Resources
Use the Resource Planner on page 181 to assign additional exercises from the program.

Alternative Assessment
Interview
Write the following measurements on the chalkboard:

    3.2 kilometers, 4,500 milligrams, 7 liters

Ask:
- Which measurement's prefix expresses a value of 1,000? Which measurement's prefix expresses a value of .001? *Answers:* kilo, milli
- What is the number of meters in 3.2 kilometers, the number of grams in 4,500 milligrams, and the number of centiliters in 7 liters? *Answers:* 3,200 meters, 4.5 grams, 700 centiliters
- About how many kilometers is 1 mile? About how many quarts is 7 liters? *Answers:* 1.6 kilometers, 7 quarts

Presentation
Ask students to illustrate each of the following concepts and give an example as they make a presentation to the class:
- Prefixes in the metric system
- Changing from a smaller unit of measure to a larger unit of measure
- Changing from a larger unit of measure to a smaller unit of measure
- Comparing customary units and metric units

Planning the Chapter

Chapter 16 • Geometry

Chapter at a Glance

SE page	Lesson	
336		Chapter Opener and Project
338	16.1	Points and Lines
340	16.2	Measuring Angles
342	16.3	Drawing Angles
343		On-The-Job Math: Physical Therapist
344	16.4	Angles in a Triangle
346	16.5	Polygons
348	16.6	Perimeter
350	16.7	Area of Squares and Rectangles
352	16.8	Area of Parallelograms
353		Using Your Calculator: Finding Perimeter and Area
354	16.9	Area of Triangles
356	16.10	Circumference of Circles
358	16.11	Area of Circles
360	16.12	Problem Solving: Subtracting to Find Area
362	16.13	Volume of Prisms
364	16.14	Volume of Cylinders
366		Chapter Review and Group Activity
368		Unit 4 Review

Learning Objectives

- Recognize points and lines.
- Use a protractor and calculate angles in a triangle.
- Identify polygons.
- Find perimeter, circumference, area, and volume.
- Solve area word problems.
- Apply angle measurement to physical therapy.

Life Skills

- Find dimensions of everyday objects.
- Use a calculator to find perimeter and area.
- Find circumference, area, and volume of everyday objects.
- Find the area of a frame or border.
- Compare estimated and actual volumes.

Communication Skills

- Explain how to measure and draw angles.
- Read a range-of-motion tool.
- Prepare questions to ask a speaker.

Workplace Skills

- Measure range of motion.
- Determine a patient's percent improvement.

PREREQUISITE SKILLS

To assess mastery of prerequisite skills, use a selection of exercises from any of the program resources referenced below. The same resources can be used to provide remediation if necessary.

	Program Resources for Review		
Skills	Student Edition	Workbook	Classroom Resource Binder
Multiplying whole numbers	Lessons 4.2, 4.3	Exercises 27, 28	Mixed Practice 44
Multiplying decimals	Lesson 10.7	Exercise 90	Review 123 Practice 129
Multiplying fractions and whole numbers	Lessons 8.1–8.3	Exercises 65–67	Reviews 88–89
Adding whole numbers	Lessons 2.2, 2.4	Exercises 11, 13	Mixed Practice 18
Subtracting whole numbers	Lessons 3.2, 3.3	Exercises 19, 20	Mixed Practice 30

Diagnostic and Placement Guide

Chapter 16 Geometry

The Percent Accuracy scores are based on the number of problems in a lesson that have been answered correctly.

Resource Planner

After diagnosing your students' needs, use the correlating Program Resources for reinforcement, reteaching, or enrichment. Additional activities for customizing lessons can be found in this guide.

KEY
Reteaching = ⤴   Reinforcement = ⬇   Enrichment = ⤴

Lessons		Percent Accuracy				Program Resources			
		50%	65%	80%	100%	Student Edition Extra Practice	Workbook Exercises	Teacher's Planning Guide	Classroom Resource Binder
16.1	Points and Lines	9	12	14	18		142	⬇ p. 193	
16.2	Measuring Angles	1	2	2	3	p. 447	143	⬇ p. 193	Visual 19 *Protractors* ⬇ Review 198 *Measuring Angles*
16.3	Drawing Angles	4	5	6	8		144	⤴ p. 194	
	On-The-Job Math: Physical Therapist	Can be used for portfolio assessment.						⤴ p. 195	⤴ On-The-Job Math 209 *Physical Therapist*
16.4	Angles in a Triangle	3	4	5	6	p. 447	145	⤴ p. 196	
16.5	Polygons	5	6	7	9		146	⬇ p. 197	
16.6	Perimeter	5	6	7	9	p. 448	147	⤴ p. 197	⬇ Review 199 *Perimeter*
16.7	Area of Squares and Rectangles	3	4	5	6	p. 448	148	⤴ p. 198	⬇ Review 200 *Area of Squares and Rectangles*
16.8	Area of Parallelograms	3	4	5	6	p. 448	149	⤴ p. 199	⬇ Review 201 *Area of Parallelograms*
16.9	Area of Triangles	5	6	7	9	p. 448	150	⬇ p. 200	⬇ Mixed Practice 207 *Area of Polygons*
16.10	Circumference of Circles	6	8	10	12	p. 449	151	⤴ p. 201	⬇ Review 202 *Circumference of Circles*
16.11	Area of Circles	6	8	10	12	p. 449	152	⤴ p. 201	⬇ Review 203 *Area of Circles* ⬇ Mixed Practice 208 *Circles* ⤴ Challenge 210 *Fixing It Up*
16.12	Problem Solving: Subtracting to Find Area	1	2	2	3		153	⤴ p. 202	⬇ Practice 206 *Problem Solving: Subtracting to Find Area*
16.13	Volume of Prisms	3	4	5	6	p. 449	154	⤴ p. 203	⬇ Review 204 *Volume of Prisms*
16.14	Volume of Cylinders	3	4	5	6	p. 449	155	⤴ p. 203	⬇ Review 205 *Volume of Cylinders*

Lessons	Percent Accuracy				Program Resources			
	50%	65%	80%	100%	Student Edition Extra Practice	Workbook Exercises	Teacher's Planning Guide	Classroom Resource Binder
Chapter 16 Review								
Vocabulary Review	4.5	5.5	7.5	9			p. 204	Vocabulary 197
	(writing is worth 4 points)							
Chapter Quiz	8	12	13	16			p. 205	Chapter Tests A, B 211, 212
Unit 4 Review	4.5	5.5	7.5	9			p. 205	Unit Test p. T13
	(critical thinking worth 4 points)							

Customizing the Chapter

Opening the Chapter
Student Edition pages 336–337

Photo Activity

The photo is of a large red cube with a cylindrical hole in the center. This is a sculpture on a plaza in front of an office building near Wall Street in New York City. Discuss the properties of a cube. A cube has six faces, all of which are squares. This cube has a hole in it, which is not characteristic of all cubes. It is an expression of the artist's concept for the sculpture. Ask students to identify everyday objects that are in the shape of a cube or a cylinder.

Words to Know

Review with students the Words to Know on page 337 of the Student Edition. To help students remember these words, have them make a labeled diagram or sketch that illustrates each one.

The following words and definitions are covered in this chapter.

angle figure formed by two rays with the same endpoint

protractor a tool used to measure angles

polygons plane figures with three or more sides; examples are *triangles*, *quadrilaterals*, *pentagons*, *hexagons*, and *octagons*

perimeter the distance around a figure

area the amount of space inside a figure

parallelogram a quadrilateral whose opposite sides are parallel; examples are *rectangles* and *squares*

circumference the distance around a circle

radius the distance from the center of a circle to its edge

space figure a three-dimensional figure that has length, width, and height

volume the amount of space inside a three-dimensional figure

Box Project

Summary: Students measure and record the lengths, perimeters, and areas of each side of a box as well as its volume.

Materials: boxes of different sizes and shapes, ruler, paper, pencils

Procedure: Students can work on this project at home or in school. Have students measure the sides of each face accurately before calculating the perimeters and areas. Then have students calculate the volumes. Ask students to summarize their results in a chart.

Display the boxes in one section of the classroom. Ask students to predict the side of the box that will have the greatest perimeter and area. Then compare their predictions with those calculated measures in the chart.

Assessment: Use the Individual Activity Rubric on page xi of this guide. Fill in the rubric with the additional information below. For this project, students should have:

- measured each side correctly.
- calculated the perimeter and area of all faces, as well as the volume of the box correctly.

Learning Objectives

Review with students the Learning Objectives on page 337 of the Student Edition before starting the chapter. Students can use the list of objectives as a learning guide. Suggest that they write the objectives in a journal or use CM6 *Goals and Self-Check* of the Classroom Resource Binder.

After each lesson, have students write an example of the skill they learned under the appropriate objective. Suggest that students use these notes as a learning guide to help them study for the chapter test.

16.1 Points and Lines
Student Edition pages 338–339

Prerequisite Skill
None

Lesson Objectives
- Identify and draw points, lines, line segments, and rays.
- Life Skill: Recognize points, lines, line segments, and rays in everyday objects.

Cooperative Group Activity

Geometry Flash Cards

Materials: index cards, pencils

Procedure: Organize the class into pairs. Give each student four index cards.

- Students write the words *point, line, line segment,* and *ray* on one side of separate index cards. On the back, the students draw a labeled picture of the figure: for example,

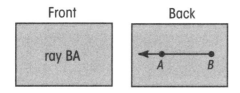

- Each pair combines its cards, shuffles them, and places them in a pile.
- One student holds up an index card with the picture of the figure facing his or her partner. The partner names the figure.
- Students alternate roles and repeat the activity until all cards have been used.
- After completing the activity, students retrieve their own cards to use as study aids.

Customizing the Activity for Individual Needs

ESL Have students draw and label pictures to help them differentiate among *lines, line segments,* and *rays*.

Learning Styles Students can:

 point out objects from the classroom that represent the basic geometric figures.

 draw representations of line, line segment, and ray on the board.

 take turns explaining the differences among meanings of point, line, line segment, and ray with a partner.

Reinforcement Activity

Have students work in pairs to create posters showing each vocabulary word and a picture of each geometric figure in everyday life. Students may find pictures in magazines. For example, points and line segments can be found in buildings. Display the posters around the classroom.

Alternative Assessment

Students can recognize basic geometric figures. Which figure represents ray CD?

Answer: c

16.2 Measuring Angles
Student Edition pages 340–341

Prerequisite Skill
- Reading whole numbers

Lesson Objective
- Use a protractor to measure an angle.

Words to Know
angle, protractor

Cooperative Group Activity

How Close Can You Get?

Materials: paper, pencils, Visual 19 *Protractors* or protractors, rulers, scissors

Procedure: Organize the class into pairs. Give each pair a ruler, a copy of Visual 19 or a protractor, and scissors.

- One partner draws a small angle with the opening facing right and estimates its measure.
- The other partner measures the angle with a protractor. The student then writes its measure inside the angle. The partners compare the estimate with the measurement.
- Students alternate roles. They repeat the activity for small and large angles with their openings facing left, right, up, and down until six angles have been drawn, estimated, and measured.

Customizing the Activity for Individual Needs

ESL To increase students' understanding of the word *degree*, say and draw its symbol. Point out that degrees are used to measure angles and temperature. Show students both a protractor and a thermometer. Write a sentence for each use, such as "The measure of angle ABC is 30°. On Monday, the temperature was 30°, and it was cold."

Learning Styles Students can:

 shade the area between the rays of an angle.

 use their arms to form an angle of a given measure to acquire a sense of the size of the angle and its orientation.

 explain to a classmate how to use a protractor; then they should have the other student repeat the explanation aloud.

Reinforcement Activity

Have students investigate the angles that the hands of a clock make. For example, the angle the hands make at 3:00 is about 90°. Students can bring in a small alarm clock with movable hands. Have them measure the angles the hands make at different times and record the results.

Alternative Assessment

Students can demonstrate which scale to use when measuring angles with a protractor.

Example: Show how to select a scale when measuring angle *ABC*.

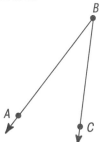

Answer: Students should place the center of the protractor's straight edge on the vertex of the angle, point *B* and then line up one ray to pass through 0° on the protractor. They use the scale that reads 0° on the first ray to read the number of degrees where the second ray crosses the protractor.

16.3 Drawing Angles
Student Edition page 342

Prerequisite Skills
- Reading whole numbers
- Measuring angles

Lesson Objective
- Use a protractor to draw an angle of a given measure.

Cooperative Group Activity

Angle-Makers

Materials: fasteners, 1-inch strips of paper, protractors

Procedure: Organize students into pairs. Give each student a fastener and two strips of paper. Write the following angle measures on the chalkboard: 45°, 105°, 85°, 15°, 170°.

- Each partner forms angle-makers by fastening the ends of the two strips of paper together.
- Partners use their angle-makers to form each angle by estimating.
- Partners switch angle-makers and use a protractor to see how close they are. They adjust the angle if necessary.
- Partners take turns picking a number between 0° and 180°, forming an estimated angle, and measuring the angle.

Customizing the Activity for Individual Needs

ESL Students can label their protractors with the word *protractor* and their rulers with the word *ruler* to help them to distinguish between the two measuring instruments. Ask students to hold up the correct instrument when you call out its name.

Learning Styles Students can:

 highlight each scale on the protractor a different color.

 use an angle-maker to form angles and then trace the angles on paper.

describe the step-by-step procedure for drawing an angle of a given measure to the class.

Enrichment Activity

Introduce the words *right, acute,* and *obtuse* to describe the size of an angle. Draw the following on the chalkboard:

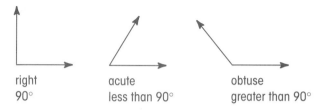

right 90°   acute less than 90°   obtuse greater than 90°

Have students draw and label each kind of angle. Then use a protractor to find the exact measurement of each angle drawn.

Alternative Assessment

Students can choose the correct drawing for a given angle.

Example: Cathy drew a 120° angle. Tell which drawing below is correct. Explain.

a) 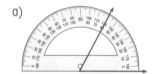   b)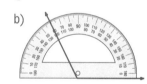

Answer: (b); explanations should include reading the scale where one ray passes through 0°.

Students should realize that since 120° is greater than 90°, you need to pass 90 on the scale.

ON-THE-JOB MATH
Physical Therapist
Student Edition page 343

Lesson Objectives
- Apply angle measurement to physical therapy.
- Communication Skill: Read a range-of-motion tool.
- Communication Skill: Prepare questions to ask a physical therapist.
- Workplace Skill: Use angle measures to find range of motion.
- Workplace Skill: Calculate a patient's percent increase in motion.

Activities
Tools of the Trade
Materials: magazines, Visual 19 *Protractors* or protractors, posters, scissors, tape, pencils
Procedure: Organize students in small groups.
- Have each group member bring in pictures from magazines of people involved in exercising.
- Have the members mark one ray on two pictures to show the different angles that parts of the body make doing the exercise or the angle the body makes with the ground.
- Students measure and label these angles.
- Groups mount their pictures on posters to make a bulletin-board display.
- Students could also write a report about the benefits of the exercises pictured on their posters.

Ask a Speaker
Materials: none
Procedure: Invite a physical therapist to speak to the class. Have students prepare questions for the therapist in advance of the visit. They should ask for information about:
- the education and training necessary for such a profession.
- the opportunities available.
- the use of angle measurement to determine the range of motion of limbs.

Act It Out
Materials: protractor, strips of cardboard, paper fasteners
Procedure: Organize the class into small groups. Provide each group with a protractor, two strips of cardboard, and a paper fastener.
- One student in each group makes an angle-maker with the cardboard and the paper fastener.
- The students use the angle-makers to measure the range of motion of the various joints: wrist, elbow, shoulder, knee, and foot. To measure the wrist, one student holds his or her hand out straight and then bends it downward as far as possible. Another student positions the angle-maker with the vertex at the bend of the wrist. Without changing the angle, the student places the angle-maker on a protractor to measure it. The range of motion of each of the joints can be measured in a similar way.
- Each student in the group measures and records the group's range of motion.
- Students may then find the mean, median, and mode for the data, and they should compare the information with other groups.

(Continued on p. 196)

(Continued from p. 195)

Practice

Have students complete On-The-Job Math Exercise 209 *Physical Therapist* from the Classroom Resource Binder.

Internet Connection

The following Web sites will give information about the job of a physical therapist:

Bureau of Labor Statistics, Occupational Outlook Handbook
   http://stats.bls.gov/oco/ocos167.htm
American Association of Physical Therapists
   http://www.apta.org

16.4 Angles in a Triangle
Student Edition pages 344–345

Prerequisite Skills
- Adding whole numbers
- Subtracting whole numbers
- Measuring angles

Lesson Objectives
- Given the measures of two angles of a triangle, find the measure of the third angle.
- Workplace Skill: Apply angle measurement to construction.

Cooperative Group Activity

Triangle Sums

Materials: protractors, paper, pencils

Procedure: Have students work in pairs. Give each student a protractor.
- Students draw three triangles of different sizes and shapes.
- Partners exchange papers and then measure the angles of each triangle. They label the angles as they are measured.
- Students find the sum of the three angles in each triangle. The sum should be 180°. Point out to students that there is always a chance for a slight error when measuring any object. Students check their angle measurements if the sum is not 180°.
- Each student draws a fourth triangle and measures two of the angles. The students find the third angle by adding the two measured angles and subtracting the sum from 180°. Then students use a protractor to check the measure of the third angle.

Customizing the Activity for Individual Needs

ESL To increase students' understanding of the word *triangle*, separate the prefix *tri-* from *angle*, and explain that *tri* means *three*. Demonstrate *tri* by using other words, accompanied by pictures, such as: *tripod*, *tricycle*, and *triathlon*. Ask students to research the dictionary for similar words.

Learning Styles Students can:

 estimate the missing angle measure by looking. Then they check the calculated result with the estimate to see if the calculated answer is reasonable.

 color each angle of a triangle a different color. Then they cut the angles apart and rearrange them to form a straight line, showing that they add up to 180°.

 make a tape recording of the steps for finding the size of the third angle of a triangle without measuring, and play it back.

Enrichment Activity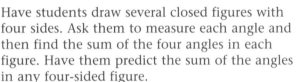

Have students draw several closed figures with four sides. Ask them to measure each angle and then find the sum of the four angles in each figure. Have them predict the sum of the angles in any four-sided figure.

Answer: 360°

Alternative Assessment

Students can explain how to find the third angle of a triangle given the measure of two angles.

Example: How can you find the third angle of a triangle if the measures of two angles are 25° and 65°?

Answer: Find the sum 25° + 65° = 90°; subtract 90° from 180°; the third angle is 90°.

16.5 Polygons
Student Edition pages 346–347

Lesson Objectives
- Identify polygons.
- Classify triangles and quadrilaterals.

Words to Know
polygons

Cooperative Group Activity

Name That Polygon!
Materials: Visual 18 *Rulers: Inch and Centimeter* or rulers, Visual 19 *Protractors* or protractors, paper, pencils

Procedure: Have students work in pairs. Provide materials for each pair.
- One student draws a polygon. He or she uses the ruler and the protractor if the polygon has special properties, for example, a square.
- The other student writes the name of the polygon.
- Students reverse roles.

Variation: One student names a polygon. The other student draws it.

Customizing the Activity for Individual Needs
ESL Help students learn polygon names by teaching them the meaning of the prefixes: *tri-* is *three*; *quad-* is *four*; *pent-* is *five*; *hex-* is *six*; *oct-* is *eight*.

Learning Styles Students can:

 draw examples of each type of polygon to help memorize the names.

 use straws on a flat surface to model each polygon.

 recite the names of the polygons, the types of triangles, and the types of quadrilaterals.

Reinforcement Activity
Write the names of the different polygons on separate index cards. Shuffle the cards and place them facedown. Have students choose cards and draw an example of each polygon on the chalkboard.

Alternative Assessment
Students can draw a polygon, given its name.

Example: Draw a pentagon.
Answer: Students draw any five-sided closed figure.

16.6 Perimeter
Student Edition pages 348–349

Prerequisite Skills
- Measuring length
- Adding measurements

Lesson Objective
- Find the perimeter of a polygon.

Word to Know
perimeter

Cooperative Group Activity

String Polygons
Materials: strings of different integer lengths: for example, 15 inches or 18 inches; rulers, tape, paper, pencils

Procedure: Organize students into groups of three or four. Give each group a piece of string, a ruler, and tape.
- Students in each group use their string to make a triangle. They sketch the figure and measure its dimensions. Then they find its perimeter.
- Groups make as many other polygons as they can with a perimeter the same length as their string. For each polygon, they make a sketch, measure the dimensions, and find its perimeter.
- Each group makes a poster to show its work.

Customizing the Activity for Individual Needs
ESL To increase students' understanding of the word *perimeter*, write *perimeter* on the chalkboard, underlining the word *rim* in its interior. Explain that the length or distance around the rim of a figure is called its perimeter. Show examples of the perimeter of polygons on the chalkboard.

Learning Styles Students can:

 make drawings of polygons and use them when solving perimeter problems.

 draw a polygon and touch each side as its perimeter is found.

 read aloud the lengths of the sides to a partner who then adds them in order to find the perimeter.

Enrichment Activity

Have students calculate the perimeter of a lot or part of the sidewalk in front of their home. Instead of using feet or yards, students can pace the perimeter and record the number of paces on each side of the lot. Have them sketch the lot or sidewalk, label the length of each side, and calculate the perimeter in paces. As an extra step, have them measure a typical pace and calculate the perimeter in feet.

Alternative Assessment

Students can draw and label a figure with a given perimeter.

Example: Draw a rectangle that has a perimeter of 20 feet.

Answer: Students draw a rectangle in which the sum of the lengths of the sides is 20 feet. The length of the opposite sides of the rectangle must be equal.

16.7 Area of Squares and Rectangles
Student Edition pages 350–351

Prerequisite Skill
- Multiplying whole numbers, decimals, and fractions

Lesson Objectives
- Find the area of squares and rectangles.
- Life Skill: Use perimeter and area to find dimensions related to a garden.

Word to Know
area

Cooperative Group Activity

Finding Area

Materials: graph paper, pencils

Procedure: Organize students into pairs. Give each student a piece of graph paper.
- One student draws a square or a rectangle on graph paper. The student finds the area in square units by counting the number of squares in the region.
- The partner finds the area by multiplying the length by the width.
- Students alternate roles and repeat the activity eight times.
- Encourage partners to compare the figures and the areas. Have them group figures with the same area together. Then they should write a statement about their observations.

Customizing the Activity for Individual Needs

ESL To increase students' understanding of the words *area*, *length*, and *width*, label a diagram. Write *long* next to *length* and *wide* next to *width*. Show students the similarity between the pairs of words and then discuss the meanings. Write *area* in the center, then shade it to show the *area*.

Learning Styles Students can:

trace the rectangle, label the two adjacent sides, and shade the interior. Then they can highlight the length in one color, the width in another color, and then they can write the formula using the same colors for each dimension.

draw and display models to show 1 square inch, 1 square foot, 1 square yard, 1 square centimeter, and 1 square meter.

read aloud the dimensions of the rectangles as they find the area.

Enrichment Activity

Provide students with graph paper. Have them find as many squares and rectangles as possible with an area of 36 square units. The sides of the quadrilaterals must be whole numbers.

Answer: 1 by 36, 2 by 18, 3 by 12, 4 by 9, and 6 by 6.

Ask students to determine which of their figures has the greatest perimeter.

Answer: 1 by 36; the perimeter is 74 inches.

Alternative Assessment

Students can explain the difference between the perimeter and the area of a figure.

Example: Explain the difference between the perimeter and area of a square that is 3 inches on each side.

Answer: Perimeter is the distance around the square. The perimeter is 12 inches. *Area* is the space inside the square. The area is 9 square inches.

16.8 Area of Parallelograms
Student Edition page 352

Prerequisite Skill
- Multiplying whole numbers, decimals, and fractions

Lesson Objective
- Find the area of a parallelogram.

Words to Know
parallelogram

Cooperative Group Activity

Double Squares
Materials: graph paper, pencils, scissors

Procedure: Organize students into groups of three. Assign a different number to each group. Start with the number 2. This will be the length of the side of a square.

- Each group draws two squares of the same size on graph paper, using the assigned number as the length of its sides. Then each group finds the total area of the two squares it drew.
- One student cuts out the two squares. Another student folds one of the squares in half diagonally and then cuts along the diagonal to form two right triangles. (See below.)

- A third student rearranges the pieces to form a parallelogram, and the group finds the area.

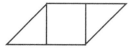

- Groups compare the area of the parallelogram to the area of the two squares from which they started. Students should recognize that the total area of the two squares is equal to the area of the parallelogram.

Customizing the Activity for Individual Needs
ESL To help students understand the words in the formula for the area of a parallelogram, have them draw a parallelogram and label the *base* and *height*. Then have them shade in the figure to represent *area*.

Learning Styles Students can:

 cross out any measurements of sides that are not used in finding the area of a parallelogram.

 trace each parallelogram and highlight only the base and height before finding the area.

 describe to a partner the steps for finding the area of a parallelogram.

Reteaching Activity

Have students find the area of a parallelogram by making it into a rectangle. Provide students with graph paper and scissors. Have them draw a parallelogram on graph paper, and then draw a height inside the parallelogram to form a triangle and a quadrilateral. Then have students cut out the parallelogram, cut off the triangle, and rearrange the pieces to form a rectangle, as shown below.

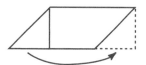

Help students realize that the base and height of the parallelogram become the length and width of the rectangle.

Alternative Assessment

Students can describe a parallelogram with a given area.

Example: Find the length and the width of a parallelogram with an area of 30 square feet.

Possible answer: The height of the parallelogram is 10 feet and the base is 3 feet.

USING YOUR CALCULATOR
Finding Perimeter and Area
Student Edition page 353

Lesson Objective
- Life Skill: Use a calculator to find the perimeter and the area of figures.

Activities

Sports Fields
Materials: reference books of sports facts, paper, pencils, calculator

Procedure: Organize students into small groups. Provide reference books of sports facts.

(Continued on p. 200)

(Continued from p. 199)

- Using reference books, students find the dimensions of different sports fields. Examples include football field, baseball field, tennis court, and boxing rink.
- Students draw a representation of the sports arena and label the dimensions.
- Using a calculator, students find the perimeter and area of the playing field for the sport they chose.
- Students may write or make a brief oral report of the sport.

Tangrams

Materials: poster board or tangrams, scissors, calculator, paper, pencils

Procedure: Organize students into small groups. Provide each group with a tangram or have students make their own from the pieces below.

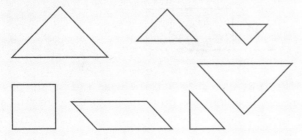

- Students arrange the pieces to form the tangram square.
- They measure the dimensions of the square to the nearest unit. Then using a calculator, they find the perimeter and area.
- Students find different arrangements of the tangram pieces. They find the perimeter and area of each arrangement. Students should notice that the area will be the same, but the perimeter will vary.

16.9 Area of Triangles
Student Edition pages 354–355

Prerequisite Skills
- Measuring length
- Multiplying whole numbers, decimals, and fractions

Lesson Objectives
- Find the area of a triangle.
- Life Skill: Compare areas of different-shaped pennants.

Cooperative Group Activity

Halving Quadrilaterals

Materials: paper, pencils, rulers

Procedure: Organize students into groups of three. Provide each group with materials.

- One student draws a square, measures its sides, and finds its area.
- Another student draws a diagonal in the square, forming two triangles. The student shades one of the triangles and finds its area using the formula $\frac{1}{2} \times$ base $\times$ height.
- The third student compares the area of the triangle with the area of the square, checking that it is $\frac{1}{2}$ the area of the square.
- Repeat the activity using rectangles and parallelograms. Be sure that students understand that to find the area of a parallelogram, they will need the height.

Customizing the Activity for Individual Needs

ESL To help students understand that the *height* is perpendicular to the base, relate it to the height of a person who is standing perpendicular to the ground.

Learning Styles Students can:

 highlight the base and height of the triangle before finding its area.

 form triangles from parallelograms.

 recite the formula for the area of a triangle into a tape recorder and play it back.

Reinforcement Activity

Have students cut out isosceles triangles from large pieces of construction paper and design pennants for their school or favorite sports team. Have them find the area and perimeter of each pennant.

Alternative Assessment

Students can find the area of a triangle when the base is outside the triangle.

Example: What is the area of the triangle below?

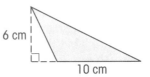

Answer: 30 square centimeters

16.10 Circumference of Circles
Student Edition pages 356–357

Prerequisite Skills
- Measuring length
- Multiplying fractions and decimals

Lesson Objectives
- Find the circumference of a circle.
- Life Skill: Apply circumference to decorating a drum.

Words to Know
circumference, radius

Cooperative Group Activity

Circling Food

Materials: food cans or other cylindrical objects, string, rulers, paper, pencils

Procedure: Organize students into small groups. Give each group string and rulers. Place the cylindrical objects in a central location.

- One group member chooses a cylindrical object and traces its base to draw a circle. Then he or she puts a dot in the center of the drawn circle as closely as possible. Another group member uses a ruler to measure the radius of the circle. Then the group calculates the circumference.
- One group member wraps the string once around the cylindrical object that the group chose and marks the place on the string. Another opens the string and measures the marked length with the ruler. The group compares this measure of circumference with the circumference that was calculated.
- Groups then choose another cylindrical object and repeat the activity.

Customizing the Activity for Individual Needs

ESL To increase students' understanding of the word *circumference*, write the word and underline *circ*. Point out that these letters also appear in the word *circle*. Draw a circle and emphasize that the *circumference* is the distance around a *circle*.

Learning Styles Students can:

 label measurements on circle drawings to help in selecting which formula to use.

 measure the circumference with string.

 tell a partner how to find the circumference of a circle.

Enrichment Activity
Have students bring in various sports balls, such as a baseball, soccer ball, basketball, or tennis ball. Have them measure the circumference of each ball and divide by 3.14 to find the approximate diameter. Have them make a chart.

Alternative Assessment
Students can estimate the circumference of a circle by using 3 for π.

Example: Find the circumference of a circle with a radius of 4 feet. Use 3 for π. Why is this an estimate?

Answer: About 24 feet; 3 is an approximate value for π.

16.11 Area of Circles
Student Edition pages 358–359

Prerequisite Skills
- Measuring length
- Multiplying decimals

Lesson Objectives
- Find the area of a circle.
- Life Skill: Find the area and circumference of a skating rink.

Cooperative Group Activity

Going Around in Circles

Materials: compasses, inch and centimeter rulers, paper, pencils

Procedure: Organize students into groups of four. Give each group a compass and a ruler.
- The first student uses the compass to draw a circle.
- The second student draws a radius from the center point to the edge.
- The third student measures and records the radius.
- The fourth student calculates and records the area.
- Students alternate roles and repeat the activity with a circle of a different size three more times. Groups exchange papers and check each other's work.

Customizing the Activity for Individual Needs

ESL To help students differentiate between the words *circumference* and *area* of a circle, have them make self-study cards. Then point out the difference between the two formulas.

$C = \pi \times 2 \times r$ $A = \pi \times r \times r$

Learning Styles Students can:

 write out the squared expression as a multiplication expression: for example, write 5×5 instead of 5^2.

 use a compass and ruler to draw circles and measure radii before calculating areas.

 take turns explaining to a partner how to find the area of a circle.

Enrichment Activity

Have students draw circles with radii of 2 units, 3 units, 4 units, and 6 units. Have them find the area of each circle. Then have them draw circles with radii of 4 units, 6 units, 8 units, and 12 units. Have them find the area of these circles and compare the results. Help students realize that when the radius is doubled, the area is four times greater.

Alternative Assessment

Students can identify the correct multiplication expression for finding the area of a circle.

Example: Which of these number expressions would you use to find the area of a circle with a radius of 4 cm?

$\pi \times 4 \times 2$ $\pi \times 4 \times 4$ $\pi \times 8 \times 8$

Answer: $\pi \times 4 \times 4$

16.12 Problem Solving: Subtracting to Find Area
Student Edition pages 360–361

Prerequisite Skills
- Measuring length
- Finding areas
- Multiplying whole numbers
- Subtracting whole numbers

Lesson Objectives
- Solve area word problems.
- Life Skill: Find the area of a frame.
- Life Skill: Draw a diagram on graph paper to find the area of a border.

Cooperative Group Activity

Border Areas

Materials: different-colored construction paper, poster board, rulers, paper, pencils, tape, scissors

Procedure: Have students work in pairs. Give each pair two different colors of construction paper.

- Each student draws and cuts out a rectangle so that one rectangle fits inside the other, leaving a border on all sides.
- One student measures the sides of each rectangle. The other student finds and records the area of each rectangle.
- One student tapes the smaller rectangle on top of the larger rectangle so that there is an equal border on all sides. The other student subtracts the area of the smaller rectangle from the area of the larger rectangle to find the area of the border.
- Students work together to create a display of their work on a poster board.

Customizing the Activity for Individual Needs

ESL Help students understand the word *border*. Have them draw a rectangle with a border. Have them shade the border and write the word *border* in the shaded area.

Learning Styles Students can:

 shade the areas that need to be found.

 cut out rectangles to model problems.

 orally explain the steps involved in finding the area of a border.

Enrichment Activity

Have students make the following drawings on graph paper. Then they should find the area of the unshaded region in each piece.

Point out that the shaded region is in a different position in each piece but that the area of the unshaded region is the same. Have students experiment with other drawings.

Alternative Assessment

Students can explain how to find the area of a border.

Example: How can you find the area of the shaded region?

Answer: Subtract the area of the smaller rectangle from the area of the larger rectangle.

16.13 Volume of Prisms
Student Edition pages 362–363

Prerequisite Skills
- Multiplying whole numbers and decimals
- Measuring length

Lesson Objectives
- Find the volume of a prism.
- Life Skill: Find the volume of a fish tank.

Words to Know
space figure, volume

Cooperative Group Activity

How Big Is the Gift?

Materials: gift boxes of various sizes, rulers, paper, pencils

Procedure: Organize students into groups of three. Give each group a ruler. Place the boxes in a central location.
- The first student chooses a box. This student measures the length, width, and height of the box.
- The second student calculates the volume of the box.
- The third student records the dimensions and volume in a chart.
- Students alternate roles and repeat the activity until all groups have found the volume of four boxes.
- Groups exchange charts and check each other's work.

Customizing the Activity for Individual Needs

ESL To help students understand the words in the formula for the volume of a rectangular prism, have them bring in small boxes and label the *length*, *width*, and *height* of each box.

Learning Styles Students can:

 highlight in a different color each dimension on the drawing of a rectangular prism. Then in those same colors, students can write the measures of each dimension in the formula for the volume.

 draw and label a sketch of the prism and relate dimension in the sketch to the formula.

 read the formula out loud while calculating the volume.

Enrichment Activity

Have students use the work-backward strategy to find the missing dimension of a rectangular prism. Give students the following example: A box has a volume of 120 cubic inches. Its length is 10 inches. Its width is 3 inches. What is its height? Demonstrate the solution: $10 \times 3 \times ? = 120$. Have students guess and check to find the missing number. Then give them similar problems to solve.

Alternative Assessment

Students can find the volume of a prism by counting cubes.

Example: Use cubes to make a prism that is 2 cubes high, 3 cubes long, and 2 cubes deep. Find the volume of the prism.

Answer: 12 cubic units

16.14 Volume of Cylinders
Student Edition pages 364–365

Prerequisite Skills
- Multiplying whole numbers and decimals
- Measuring length

Lesson Objectives
- Find the volume of a cylinder.
- Life Skill: Find the volume of a cylindrical water tank.

Cooperative Group Activity

Fill It Up!

Materials: cylindrical-shaped beverage mugs, rulers, paper, pencils

Procedure: Before starting the activity, display all mugs. Have students predict which mug has the greatest volume. List the mugs in order from least to greatest on the chalkboard. Title the list *Predictions*. Then organize the class into pairs. Give each pair a ruler.

- One partner chooses a mug. This student measures the radius and height of the mug.
- The other partner finds the volume of the mug using the formula.
- The students then record their results on paper.
- Pairs of students exchange mugs and repeat the activity several times.
- The class lists each mug and its volume on the chalkboard from least to greatest. Title this list *Actual Volumes*. Students check to see if their predictions were correct.

Customizing the Activity for Individual Needs

ESL Have students bring in an empty cylindrical oatmeal box or soda can and label it with the word *cylinder*. Have them place the word *height* vertically on the box or can and $\pi \times (\text{radius})^2$ on the top of the box.

Learning Styles Students can:

 highlight in a different color the radius and height on the drawing of a cylinder. Then using the same colors, they can write these measures in the formula for volume.

 use construction paper to make a model of a cylinder and label its dimensions. Be sure that students include a top and bottom for their cylinder.

 read the formula out loud while calculating the volume.

Enrichment Activity

Have students guess and check to find the missing dimension of a cylinder. Give students the following problem: The volume of a cylinder is 282.6 cubic inches. Its radius is 3 inches. What is its height? Students can write the values they know in the formula: $3.14 \times 3 \times 3 \times ? = 282.6$. Then they can use a calculator to guess and finally check to find the missing number.

Alternative Assessment

Students can demonstrate how to use the formula to find the volume of a cylinder.

Example: The bottom of a can has a diameter of 4 inches. The height is 6 inches. The formula for the volume of a cylinder is given below.

Volume = $\pi \times (\text{radius})^2 \times \text{height}$

Write the formula using the numbers you need to find the volume of the can. Use 3.14 for π.

Answer: Volume = $3.14 \times (2)^2 \times 6$

Closing the Chapter
Student Edition pages 366–367

Chapter Vocabulary

Review with the students the Words to Know on page 337 of the Student Edition. Then have students quiz each other in pairs.

Have students copy and complete the Vocabulary Review questions on page 366 of the Student Edition.

For more vocabulary practice, have them complete Vocabulary Exercise 197 *Geometry* from the Classroom Resource Binder.

Test Tips

Have pairs of students explain how to solve a problem by using one of the test tips.

Learning Objectives

Have students review CM6 *Goals and Self-Check* from the Classroom Resource Binder. They can check off the goal they have reached. Note that each section of the quiz corresponds to a Learning Objective.

Group Activity

Summary: Students estimate and then calculate the volumes of various boxes and cans.

Materials: boxes, cans, paper, pencils

Procedure: Students can use a ruler to measure those dimensions they need to calculate the volume of a can or a box. Have them compare their estimates with the actual results and prepare a report of their findings.

Assessment: Use the Group Activity Rubric on page xii of this guide. Fill in the rubric with the additional information below. For this project, students should have:

- made a reasonable estimate of the volume of each figure.
- calculated the volume of each figure correctly.

Assessing the Chapter

Traditional Assessment

Chapter Quiz
The Chapter Quiz on pages 366–367 of the Student Edition can be used as either an open- or closed-book test, or as homework. The quiz can be used to identify concepts in the chapter that students need to review and practice.

Chapter Tests
Use Chapter Test A Exercise 211 and Chapter Test B Exercise 212 *Geometry* from the Classroom Resource Binder to further assess mastery of chapter concepts.

Additional Resources
Use the Resource Planner on pages 191–192 of this guide to assign additional exercises from the program.

RELATED MATERIALS See the unit overview page for other Globe Fearon books that can be used to enrich and extend the material in this unit.

Alternative Assessment

Application
Display a rectangular prism and a cylinder: for example, a tissue box and a can. Provide students with protractors and rulers. Have students:
- measure an angle of the rectangular prism.
- find the area of one face of the prism.
- find the volume of the prism.
- use 3 for π to estimate the circumference of the circular base of the cylinder.
- find the area of the circular base of the cylinder.
- find the volume of the cylinder.

Performance-Based
Provide students with rulers and protractors. Draw the following parallelogram on the chalkboard. Note that angle *ABC* should be about 125°.

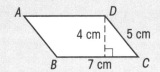

Then have students:
- name a point and a line segment.
- measure angle BCD.
- identify the polygon.
- find the perimeter of the polygon.
- find the area of the polygon.

Unit Assessment
This is the last chapter in Unit 4, *Measurement and Geometry*. To assess cumulative knowledge and to provide standardized-test practice, administer both the practice test on page 368 of the Student Edition and the Unit 4 Cumulative Test T13 from the Classroom Resource Binder. These tests are in multiple-choice format. A scantron sheet is provided on page T2 of the Classroom Resource Binder.

Unit Overview

Unit 5 ▶ Algebra

CHAPTER 17	PORTFOLIO PROJECT	MATH IN YOUR LIFE	USING YOUR CALCULATOR	PROBLEM SOLVING
Integers	Integer Search	Wind Chill Temperature	Keying In Integers	Using Integers

CHAPTER 18	PORTFOLIO PROJECT	ON-THE-JOB MATH	USING YOUR CALCULATOR	PROBLEM SOLVING
Algebra	Equation	Computer Programmer	Checking Solutions to Equations	Using a One-Step Equation Using a Two-Step Equation

RELATED MATERIALS

These are some of the Globe Fearon books that can be used to enrich and extend the material in this unit.

Practice & Remediation ▶

Pacemaker® Pre-Algebra
Prepare your students to take on algebra with this complete, accessible pre-algebra curriculum.

Test Preparation ▶

Basic Algebra
Reinforce algebra content and concepts with this manageable, focused program.

Real-World Math ▶

Buying with Sense
Teach consumer math skills needed for successful independent living through this real-world activity driven program.

Planning the Chapter

Chapter 17 • Integers

Chapter at a Glance

SE page	Lesson
370	Chapter Opener and Project
372	17.1 What Is an Integer?
374	17.2 Adding Integers with Like Signs
376	17.3 Adding Integers with Unlike Signs
378	17.4 Subtracting Integers
381	Using Your Calculator: Keying In Integers
382	17.5 Multiplying Integers
385	Math in Your Life: Wind Chill Temperature
386	17.6 Dividing Integers
388	17.7 Problem Solving: Using Integers
390	Chapter Review and Group Activity

Learning Objectives
- Identify and write integers.
- Add integers with like signs.
- Add integers with unlike signs.
- Subtract integers.
- Multiply integers.
- Divide integers.
- Solve problems using integers.
- Apply integers to wind chill temperatures.

Life Skills
- Apply integers to elevation, football plays, the stock market, and temperature.
- Find the wind chill temperature.
- Use a calculator to add or subtract integers.
- Draw a diagram.

Communication Skills
- Read integers in magazines, newspapers, and an atlas.
- Interpret a wind chill table.
- Keep a written record of wind chill temperatures.
- Explain how negative integers are used in real life.
- Explain the effect of wind chill.

PREREQUISITE SKILLS
To assess mastery of prerequisite skills, use a selection of exercises from any of the program resources referenced below. The same resources can be used to provide remediation if necessary.

Skills	Program Resources for Review		
	Student Edition	Workbook	Classroom Resource Binder
Ordering whole numbers	Lesson 1.7	Exercise 7	Practice 6 Mixed Practice 8
Adding and subtracting whole numbers	Lessons 2.2, 3.2	Exercises 11, 19	Mixed Practice 18, 30
Multiplying and dividing whole numbers	Lessons 4.2, 5.2	Exercises 27, 36	Mixed Practice 44, 59, 60

Diagnostic and Placement Guide

Chapter 17 Integers

The Percent Accuracy scores are based on the number of problems in a lesson that have been answered correctly.

Resource Planner

After diagnosing your students' needs, use the correlating Program Resources for reinforcement, reteaching, or enrichment. Additional activities for customizing lessons can be found in this guide.

KEY
Reteaching = ⤺   Reinforcement = ⬇   Enrichment = ⤻

Lessons		Percent Accuracy				Program Resources			
		50%	65%	80%	100%	Student Edition Extra Practice ⬇	Workbook Exercises ⬇	Teacher's Planning Guide	Classroom Resource Binder
17.1	What Is an Integer?	11	14	18	22		156	⤺ p. 209	Visual 20 *Number Lines: Integers*
17.2	Adding Integers with Like Signs	9	12	14	18	p. 450	157	⤻ p. 210	Visual 20 *Number Lines: Integers*; ⬇ Review 214 *Adding Integers with Like Signs*
17.3	Adding Integers with Unlike Signs	15	20	24	30	p. 450	158	⬇ p. 211	Visual 20 *Number Lines: Integers*; ⬇ Review 215 *Adding Integers with Unlike Signs*
17.4	Subtracting Integers	15	20	24	30	p. 450	159	⤺ p. 211	⬇ Review 216 *Subtracting Integers*; ⤻ Challenge 222 *The Stock Market*
17.5	Multiplying Integers	18	23	29	36	p. 451	160	⬇ p. 213	⬇ Review 217 *Multiplying Integers*
	Math in Your Life: Wind Chill Temperature	Can be used for portfolio assessment.						⤺ p. 213	⤻ Math in Your Life 221 *Wind Chill Temperature*
17.6	Dividing Integers	18	23	29	36	p. 451	161	⬇ p. 214	⬇ Review 218 *Dividing Integers*; ⬇ Mixed Practice 220 *Adding, Subtracting, Multiplying and Dividing Integers*
17.7	Problem Solving: Using Integers	1	2	2	3	p. 451	162	⤺ p. 215	Visual 20 *Number Lines: Integers*; ⬇ Practice 219 *Problem Solving: Using Integers*
Chapter 17 Review									
	Vocabulary Review	4	5.5	6.5	8 (writing is worth 4 points)			⬇ p. 216	⬇ Vocabulary 213
	Chapter Quiz	20	26	32	40			⬇ p. 216	⬇ Chapter Test A, B 223, 224

Customizing the Chapter

Opening the Chapter
Student Edition pages 370–371

Photo Activity
Discuss the meaning of height in terms of feet above and below sea level. Relate this concept to a vertical number line with zero as the point that represents sea level. Help students to understand negative numbers. Name different integers. Have students tell whether the integers represent heights that are below sea level or above sea level.

Words to Know
Review with the students the Words to Know on page 371 of the Student Edition. To help students remember these words, have them draw and label a number line.

The following words and definitions are covered in this chapter.

integers numbers in the set {..., $^-3$, $^-2$, $^-1$, 0, $^+1$, $^+2$, $^+3$, ...}

positive integers integers to the right of zero on the number line

negative integers integers to the left of zero on the number line

opposites two numbers that are the same distance from zero on the number line but are on opposite sides of zero

Integer Search Project
Summary: Students find real-life examples of integers: for example, temperature, the stock market, money, height/depth, and sports.

Materials: newspapers, magazines, or any print medium; adding machine tape; paper; pencils

Procedure: Students can complete this project in class with all materials provided. Or, students can keep a journal throughout this chapter and cite examples found during their regular daily routine. Students can also make a class number line using adding machine tape and marking the integers they have found. Be sure that students understand that integers are whole numbers and their opposites. Negative fractions or negative decimals are not integers.

Assessment: Use the Individual Activity Rubric on page xi of this guide. Fill in the rubric with the additional information below. For this project, students should have:
- found at least five integers, one per topic.
- described what each integer represents.

Learning Objectives
Review with the students the Learning Objectives on page 371 of the Student Edition before starting the chapter. Students can use the list of objectives as a learning guide. Suggest that they write the objectives in a journal or use CM6 *Goals and Self-Check* from the Classroom Resource Binder.

After each lesson, have students write an example of the skill they learned under the appropriate objective. Suggest that students use these notes as a learning guide to help them study for the chapter test.

17.1 What Is an Integer?
Student Edition pages 372–373

Prerequisite Skills
- Writing whole numbers on a number line
- Ordering whole numbers

Lesson Objectives
- Locate an integer on a number line.
- Find the opposite of an integer.
- Write an integer for a real-life situation.

Words to Know
integers, positive integers, negative integers, opposites

Cooperative Group Activity

Build a Number Line
Materials: self-stick notes, paper, pencils

Procedure: Make one self-stick note per student and one for yourself. Write an integer on each note. Prepare an equal number of positive and negative integers. Mix up the order of the numbers. Organize students into two equal teams. Give each student one self-stick note. Keep zero for yourself. Draw a blank number line on the chalkboard. Place zero in the middle.

- One student from each team comes to the chalkboard to place their self-stick notes in the appropriate place on the number line. Students should leave room for missing numbers.
- As soon as the team member sits down, the next one on the team should go up to the chalkboard. Repeat until each team member has gone. The first team done wins.

Chapter **17**

- After the number line has been built, one team member names an integer. A member of the other team names its opposite. That team member then names another integer and a different member of the other team names its opposite, and so on.

Customizing the Activity for Individual Needs

ESL To help students recognize words associated with *negative* and *positive* in relation to integers, write *negative* and its symbol on the back of one index card. Do the same for the word *positive*. Write examples of both integers on the chalkboard, as well as some words such as: *gain, loss, above,* and *below*. As you point to each one, the students should raise the correct card.

Learning Styles Students can:

 highlight positive integers in one color and negative integers in another color on the number line or in text.

 use Visual 20 *Number Lines: Integers* as a reference to leave on the desk or in the notebook.

 count integers aloud starting from zero to find the position of integer on the number line.

Reinforcement Activity

Provide students with copies of Visual 20 *Number Lines: Integers* or a number line. Have students begin by pointing with a pencil to zero. Then have them follow your directions to locate integers on the number line. Examples of directions might be, "Go 3 integers right, then 6 integers left. What integer do you land on?"

Alternative Assessment

Students can order a consecutive group of integer cards and identify a missing integer.

Example: Order these integers. What integer is missing?

Answer: ⁻3, ⁻2, 0, 1, 2, 3; ⁻1 is missing.

17.2 Adding Integers with Like Signs
Student Edition pages 374–375

Prerequisite Skill
- Adding whole numbers on the number line

Lesson Objectives
- Add integers with like signs.
- Life Skill: Use integers to represent distances up and down a mountain.

Cooperative Group Activity

I.O.U.

Materials: Visual 20 *Number Lines: Integers* or number line, index cards, paper, pencils

Procedure: Organize students into groups of four or five. Give each group 20 index cards. Explain that an I.O.U. is a promise to pay money owed. When a person pays an I.O.U., he or she returns money.

- For each number from 1 to 20, groups write separate I.O.U. cards, such as, *I owe you $3*.
- Groups shuffle their sets of cards and group members take turns choosing two cards at a time. They use the cards to form an addition problem involving negative integers. For example, *I owe you $3 and I owe you $9*.
- Students write a number expression for the word problem and use a number line to solve; for the example, ⁻3 + ⁻9 = ⁻12.
- Students explain what the answer means; for the example, *Now, I owe you $12*.
- Groups can check each other's work.
- As students develop an understanding of addition of negative integers, they can choose three cards at a time, or make new cards with larger numbers.
- Groups should hand in their sets of index cards for use with the next lesson.

Customizing the Activity for Individual Needs

ESL To help students understand the meaning of I.O.U., point out that it stands for the words *I owe you*. Have one student say an I.O.U. phrase. Another student can use play money to pay the I.O.U.

Learning Styles Students can:

 highlight the sign of the integers. Then automatically write it as part of the answer.

 make jumps with a finger on a number line to add integers with like signs.

count aloud spaces on a number line while adding integers with like signs.

Enrichment Activity

A deep-sea diving ship can dive more than 6 miles below the surface to explore ocean depths. How many feet below the surface would a ship be if it went 5 miles below the surface and then climbed up 2,640 feet? (1 mile = 5,280 feet)

Answer: 23,760 feet below the surface

Alternative Assessment

Students can give an oral response to a basic integer addition problem.

Example: What is the sum of ⁻3 and ⁻5?

Answer: ⁻8

17.3 Adding Integers with Unlike Signs *Student Edition pages 376–377*

Prerequisite Skill
- Adding whole numbers

Lesson Objectives
- Add integers with unlike signs.
- Life Skill: Use addition of integers to describe football plays.

Cooperative Group Activity

Payday

Materials: I.O.U. index cards from the preceding activity, index cards, pencils, paper

Procedure: Organize students into groups of four or five. Give each group 20 new index cards and one set of 20 I.O.U. cards from the preceding activity.

- For each number from 1 to 20, groups write separate payday cards, such as, *I earn $5*.
- Groups combine both sets of cards and shuffle them. Group members take turns choosing two cards at a time. They use the cards to form an addition problem involving integers; for example, *I earn $5 and I owe you $12*.
- Students write a number expression for the word problem and use a number line to solve for the example, ⁺5 + ⁻12 = ⁻7.
- Students explain what the answer means; for the example, *Now I owe you $7*.

- Students keep an ongoing record of their gains and losses. Once they have a balance of $30, they complete the activity and help other group members reach the same goal.

Customizing the Activity for Individual Needs

ESL To help students understand the difference between *owing* and *earning* money, have them work in pairs to role-play each. Pairs take turns picking a card containing a payday or an I.O.U. phrase and act it out. The rest of the group guesses which word describes the play.

Learning Styles Students can:

 use one color to associate a move to the *left* on the number line with a *negative* integer and another color to associate a move to the *right* on the number line with a *positive* integer.

 use two different-colored counters to model positive and negative integers.

 say aloud the direction and the number of spaces to move on a number line to add integers with unlike signs.

Reinforcement Activity

Provide a set of negative integer cards and a set of positive integer cards. Have students select one card from each pile and tell if the sum will be negative or positive. Students can then use a number line to find the actual sum.

Alternative Assessment

Students can explain how to add two integers with unlike signs on a number line.

Example: How would you add ⁺6 and ⁻9 on a number line? What is the sum?

Answer: Begin at 0. Move 6 spaces to the right. Then move 9 spaces to the left. The sum is ⁻3.

17.4 Subtracting Integers *Student Edition pages 378–380*

Prerequisite Skill
- Adding integers

Lesson Objectives
- Subtract integers.
- Life Skill: Use subtraction of integers to describe changes in depth.

Cooperative Group Activity

Spinner Subtraction

Materials: Visual 17 *Spinners* or spinners, Visual 20 *Number Lines: Integers* or number lines, paper, pencils

Procedure: Organize students into groups of three. Provide each group with materials.

- Group members work together to make the following spinners from Visual 17.

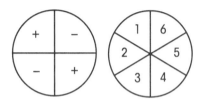

- One student spins to find a number without a sign. Another student spins the second spinner to determine the sign for the number. The two students spin again for the second number and sign. The third student writes a subtraction problem for the integers spun; for example, ⁻3 – ⁻5.
- Students work together using Visual 20 or a number line to find the difference. They then share their work with the class.

Customizing the Activity for Individual Needs

ESL Help students to distinguish between the *positive* and *negative* symbols for integers and the *addition* and *subtraction* symbols for the operations. Have students work together in pairs to translate number expressions on index cards.

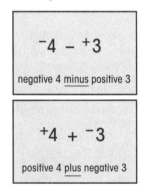

Learning Styles Students can:

 circle or highlight the signs that need to be changed.

 manipulate integer cards and the operation symbols + and – to show how to change subtraction to addition of integers.

 say *positive* or *negative* while changing signs in problems.

Enrichment Activity

Have students work in groups to model problems to answer these questions. Encourage students to try several examples and determine a general rule for each question.

- When you subtract a positive number, is the difference greater or less than the number you subtracted from?
- When you subtract a negative number, is the difference greater or less than the number you subtracted from?

Answers: less than; greater than

Alternative Assessment

Students can explain how to subtract an integer.

Example: How would you solve this problem?

⁻6 – ⁻8 = ?

Answer: Change negative 8 to positive 8. Change the subtraction to addition, then add; 2

USING YOUR CALCULATOR
Keying In Integers
Student Edition page 381

Lesson Objective
- Life Skill: Use a calculator to add and subtract integers.

Activity
Did I Add or Subtract?
Materials: a set of integer cards from ⁻10 to ⁺10, calculator

Procedure: Have students work in pairs. Have them mix a set of integer cards from ⁻10 to ⁺10 and place them facedown on a desk or table.

- The first player selects two cards and turns them faceup; for example, ⁻3 and ⁻5. He or she uses the calculator to add or subtract the numbers, then tells the other player the result; for the example, ⁺2.
- The other player then uses the calculator and trial and error to discover whether the integers were added or subtracted and in what order; for this example, ⁻3 – ⁻5 = ⁺2.
- Players then reverse roles.
- Points are awarded for correct discoveries. The first player to receive 10 points wins a round.

212 • Chapter 17

17.5 Multiplying Integers
Student Edition pages 382–384

Prerequisite Skill
- Multiplying whole numbers

Lesson Objective
- Multiply integers with like and unlike signs.

Cooperative Group Activity

Multiplication Toss

Materials: blank number cubes, markers, paper, pencils

Procedure: Organize students into groups of four. Give each group two blank number cubes.
- Each group prepares the number cubes by writing ⁻4 to ⁻9 on one cube and ⁺4 to ⁺9 on the other cube.
- One student tosses both number cubes. Another student identifies whether the product will be positive or negative. The third student finds the product. The fourth student writes the completed multiplication sentence on a piece of paper.
- Students change roles on each toss.
- After about 15 minutes, groups exchange papers to check each other's work.

Customizing the Activity for Individual Needs

ESL To help students understand the meaning of the words *different* and *same,* have them prepare the following to use as a reference.

Learning Styles Students can:

 circle or highlight the integer signs to help decide on the sign of the product.

 model the multiplication with integer cards.

 read integers aloud, stressing whether they are negative or positive integers.

Reinforcement Activity

Mix a set of integer cards from ⁻9 to ⁺9, without zero. Pick two at random and hold them up. Have students respond orally, telling you the product of the two integers selected. Make sure that students state the sign of the product.

Alternative Assessment

Students can identify the sign of the product in a multiplication problem.

Example: What is the sign in the product of a negative integer times negative integer?

Answer: positive

MATH IN YOUR LIFE
Wind Chill Temperature
Student Edition page 385

Lesson Objectives
- Apply integers to wind chill temperatures.
- Life Skill: Read a wind chill table to determine the wind chill temperature.
- Communication Skill: Keep a record of wind chill temperatures.
- Communication Skill: Explain the effect of wind chill.

Activities

Tools of the Trade

Materials: weather sections from a local daily newspaper for a period of one week and a wind chill table or an almanac, paper, pencils

Procedure: Organize students into small groups. Provide each group with the appropriate weather information. Have students:
- record the daily high and low temperatures.
- record the daily average wind speed.
- use the table on page 385 of the Student Edition (or a similar table from an almanac or encyclopedia) to find the approximate wind chill temperature for both the high and low temperatures.
- keep a record of the wind chill temperatures for the week.

Look It Up

Materials: reference books, paper, pencils

Procedure: Have students do research on wind chill temperatures. Have them report on
- the effects of wind chill on the body.
- how to dress for wind chill temperatures.

Act It Out

Materials: none

Procedure: Organize students into pairs. Present the following scenario for students to role-play.
- Trevor is planning to hike in the park.

(Continued on p. 214)

(Continued from p. 213)
- The weather reporter announces the temperature is 30° F; the winds are expected to reach 20 mph by this afternoon.
- Explain to Trevor how wind chill makes the temperature feel colder.
- Trevor must decide how he will dress to protect himself.

Practice

Have students complete the Math in Your Life Exercise 221: *Wind Chill Temperature* from the Classroom Resource Binder.

Internet Connection

Students can calculate wind chill temperatures and learn more about wind chill at the following Web sites:

The Annenberg/CPB Project Exhibit Collection, Weather
> http://www.learner.org/exhibits/weather/act_windchill/

Determining Wind Chill Temps
> http://www.lakevermilion.com/windchill/index.html

17.6 Dividing Integers
Student Edition pages 386–387

Prerequisite Skill
- Dividing whole numbers

Lesson Objective
- Divide integers with like and unlike signs.

Cooperative Group Activity

Fact Families
Materials: blank number cubes, markers, paper, pencils

Procedure: Review related multiplication and division facts for whole numbers. Write the following on the chalkboard:

   Since $7 \times 4 = 28$ then $28 \div 4 = 7$

then show how this can be applied to integers:

   Since $^+7 \times {}^-4 = {}^-28$ then $^-28 \div {}^-4 = {}^+7$

Organize students into groups of three. Give each group two blank number cubes.

- Each group prepares the number cubes by writing $^-4$ to $^-9$ on one cube and $^+4$ to $^+9$ on the other cube.
- One student tosses both number cubes. Another student writes the multiplication problem on paper and finds the product. A third student writes the related division sentence.
- Students change roles on each toss.
- After about 15 minutes, groups exchange papers to check each other's work.

Customizing the Activity for Individual Needs

ESL To help students understand the words *different* and *same* as they relate to division of integers, have students write the following multiplication problem and its inverse on index cards, with the proper labels.

different	negative	same	positive
$^+7 \times {}^-4$	$= {}^-28$	$^-28 \div {}^-4$	$= {}^+7$

Learning Styles   Students can:

 circle or highlight the signs in the problem to help determine the sign of the quotient.

 use integer cards to form related multiplication and division problems.

 read facts aloud to help decide the sign of the quotient.

Reinforcement Activity

Provide groups of students with a set of integer cards: $^+2, {}^-2, {}^+3, {}^-3, {}^+4, {}^-4, {}^+6, {}^-6, {}^+8, {}^-8, {}^+12,$ and $^-12$. Have them mix the cards and place them facedown on the table. Write $^-24$ on the board. Have students pick a card and divide $^-24$ by the integer on the card. Repeat the activity using $^+24$ as the dividend.

Alternative Assessment

Students can tell why a quotient is positive or negative.

Example: Which of these problems will have negative quotients? Which will have positive quotients? Why?

(a) $^-12 \div {}^+4$ (b) $^+12 \div {}^+4$
(c) $^-12 \div {}^-4$ (d) $^+12 \div {}^-4$

Answers: (a) and (d) will have negative quotients because the signs are different; (b) and (c) will have positive quotients because the signs are the same.

17.7 Problem Solving: Using Integers *Student Edition pages 388–389*

Prerequisite Skills
- Adding and subtracting integers with like signs
- Multiplying and dividing integers

Lesson Objectives
- Solve word problems using integers.
- Life Skill: Draw a diagram to solve word problems involving integers.

Cooperative Group Activity

The Thermometer
Materials: copies of a thermometer, paper, pencils

Procedure: Explain to students that they can use the thermometer as a number line. Illustrate that temperatures above zero would be positive integers and temperatures below zero would be negative integers. Organize students into groups of four. Provide each group with a copy of a thermometer.
- Group members work together to write four word problems about the thermometer.
- Groups exchange problems to solve. Students can use the copy of the thermometer to model each problem.
- The class reviews the problems and the solutions as a whole.

Customizing the Activity for Individual Needs
ESL To help students understand *above* and *below* zero, and an *increase* or *decrease* in temperature, have students work in pairs with a demonstration thermometer. Ask one partner to locate and mark a temperature on the thermometer, stating if it is *above* or *below* zero. The other partner then locates a new temperature on the same thermometer, stating if it shows an *increase* or a *decrease* in temperature.

Learning Styles Students can:

 draw diagrams or number lines to solve integer word problems.

 trace the movements described in a problem on a number line as they read the problem.

 take turns reading each problem aloud to a partner. Decide on the clue words and tell what operation to use to solve it.

Enrichment Activity

Students can create word problems for classmates to solve based on real-life situations involving addition and subtraction of integers: profit and loss, above and below sea level, points gained or lost in a game, rise and fall of temperatures above and below 0°.

Alternative Assessment

Students can use a number line to illustrate and solve a word problem using integers.

Example: Jenna owes Paul $5. She earns $3. Use a number line to find Jenna's balance.

Answer: Jenna still owes Paul $2.

Chapter 17

Closing the Chapter
Student Edition pages 390–391

Chapter Vocabulary
Review with the students the Words to Know on page 371 of the Student Edition. Then have students quiz each other in pairs.

Have students copy and complete the Vocabulary Review questions on page 390 of the Student Edition.

For more vocabulary practice, have them complete Vocabulary Exercise 213 *Integers* from the Classroom Resource Binder.

Test Tips
Have pairs of students make flash cards of the test tips to help them remember the rules for addition, subtraction, multiplication, and division of integers.

Learning Objectives
Have students review CM6 *Goals and Self-Check* from the Classroom Resource Binder. They can check off the goal they have reached. Note that each section of the quiz corresponds to a Learning Objective.

Group Activity
Summary: Students find and compare elevations above and below sea level, label a diagram of the highest to lowest places, and find the differences in their elevations.

Materials: atlases, copies of an elevation chart, pencils, paper

Procedure: Organize students into groups of four. Provide each group with an atlas and a copy of an elevation chart. Have students decide which continent they will use. Then one student looks up elevations for five locations on their continent. The next student puts the names of the places on the chart. A third group member labels the correct elevations. The fourth member finds the differences: highest to lowest, second highest to second lowest, and so on.

Assessment: Use the Group Activity Rubric on page *xii* of this guide. Fill in the rubric with the additional information below. For this activity, students should have:
- recorded the information from atlases correctly as integers.
- found the differences between elevations correctly.

RELATED MATERIALS See the unit overview page for other Globe Fearon books that can be used to enrich and extend the material in this unit.

Assessing the Chapter

Traditional Assessment
Chapter Quiz
The Chapter Quiz on pages 390–391 of the Student Edition can be used as either an open-book or closed-book test, or as homework. Use the quiz to identify concepts in the chapter that students need to review and practice.

Chapter Tests
Use Chapter Test A Exercise 223 and Chapter Test B Exercises 224: *Integers* from the Classroom Resource Binder to further assess mastery of chapter concepts.

Additional Resources
Use the Resource Planner on page 208 of this guide to assign additional exercises from the program.

Alternative Assessment
Interview
Write these integers on a sheet of paper: $^+8$, $^-8$, $^+2$, $^-2$. Ask:
- Which numbers would you add to get a sum of $^+10$?
- Which numbers would you add to get a sum of $^+6$?
- Which numbers would you subtract to get $^-10$?
- Which numbers would you multiply to get $^+16$?
- Which numbers would you divide to get $^-4$?

Real-Life Connection
Ask groups of students to brainstorm other real-life situations that could be described using positive and negative integers. Students:
- explain how the integers would be used.
- tell what clue words would be used to denote positive and negative integers.
- write one word problem and its solution, each using the scenario they selected.

Planning the Chapter

Chapter 18 • Algebra

Chapter at a Glance

SE page	Lesson
392	Chapter Opener and Project
394	18.1 What Is an Equation?
396	18.2 Using Parentheses
397	On-The-Job Math: Computer Programmer
398	18.3 Order of Operations
400	18.4 Solving Equations with Addition and Subtraction
402	18.5 Solving Equations with Multiplication and Division
404	18.6 Problem Solving: Using a One-Step Equation
406	18.7 Solving Equations with More Than One Operation
409	Using Your Calculator: Checking Solutions to Equations
410	18.8 Problem Solving: Using a Two-Step Equation
412	Chapter Review and Group Activity
414	Unit Review

Learning Objectives

- Identify solutions to equations.
- Simplify expressions with parentheses and using the order of operations.
- Solve equations with one operation.
- Solve equations with more than one operation.
- Solve problems using one-step equations and two-step equations.
- Apply algebra to computer programming.

Life Skills

- Find the cost of multiple items.
- Use algebra to represent everyday situations.
- Find distance, rate, or time.
- Calculate taxi fares and rental costs.
- Use a calculator to check solutions to equations.
- Make a table.

Communication Skills

- Explain how equations are used.
- Explain the steps in a computer program.
- Write about the cost of renting an item.

Workplace Skills

- Understand a computer program written in Basic.
- Write a computer program in Basic.

PREREQUISITE SKILLS

To assess mastery of prerequisite skills, use a selection of exercises from any of the program resources referenced below. The same resources can be used to provide remediation if necessary.

Program Resources for Review

Skills	Student Edition	Workbook	Classroom Resource Binder
Adding and subtracting whole numbers	Lessons 2.2, 3.2	Exercises 11, 19	Mixed Practice 18, 30
Multiplying and dividing whole numbers	Lessons 4.2, 5.2	Exercises 27, 36	Mixed Practice 44, 60

Diagnostic and Placement Guide

**Chapter 18
Algebra**

The Percent Accuracy scores are based on the number of problems in a lesson that have been answered correctly.

Resource Planner

After diagnosing your students' needs, use the correlating Program Resources for reinforcement, reteaching, or enrichment. Additional activities for customizing lessons can be found in this guide.

KEY
Reteaching = ↶   Reinforcement = ↓   Enrichment = ↷

Lessons	Percent Accuracy				Program Resources			
	50%	65%	80%	100%	↓ Student Edition • Extra Practice	↓ Workbook Exercises	Teacher's Planning Guide	Classroom Resource Binder
18.1 What Is an Equation?	9	12	14	18		163	↶ p. 219	Visual 21 *Balance*
18.2 Using Parentheses	6	8	10	12	p. 452	164	↓ p. 220	
On-The-Job Math: Computer Programmer	*Can be used for portfolio assessment.*						↶ p. 221	↶ On-The-Job Math 233 *Computer Programmer*
18.3 Order of Operations	12	16	19	24	p. 452	165	↶ p. 221	↓ Review 226 *Order of Operations*
18.4 Solving Equations with Addition and Subtraction	11	14	17	21	p. 452	166	↓ p. 222	↓ Review 227 *Solving Equations with Addition and Subtraction*
18.5 Solving Equations with Multiplication and Division	9	12	14	18	p. 452	167	↶ p. 223	↓ Review 228 *Solving Equations with Multiplication and Division* ↓ Mixed Practice 232 *Solving Equations*
18.6 Problem Solving: Using a One-Step Equation	1	2	2	3	p. 453	168	↓ p. 223	↓ Practice 230 *Problem Solving: Using a One-Step Equation*
18.7 Solving Equations with More Than One Operation	19	25	30	38	p. 453	169	↶ p. 224	↓ Review 229 *Solving Equations with More than One Operation* ↓ Mixed Practice 232 *Solving Equations*
18.8 Problem Solving: Using a Two-Step Equation	1	2	2	3	p. 453	170	↶ p. 225	↓ Practice 231 *Problem Solving: Using a Two-Step Equation* ↶ Challenge 234 *Where Are You Going? What Will It Cost?*
Chapter 18 Review								
Vocabulary Review	5	6.5	8	10			↓ p. 226	↓ Vocabulary 225
	(writing is worth 4 points)							
Chapter Quiz	10	13	16	20			↓ p. 227	↓ Chapter Test A, B 235, 236
Unit 5 Review	4	6.5	8	10			↓ p. 227	↓ Unit Test p. T15
	(critical thinking is worth 4 points)							

Customizing the Chapter

Opening the Chapter
Student Edition pages 392–393

Photo Activity
Discuss how the acrobat in the photo can add or subtract the same number of circles and still be balanced. Relate this idea to a pan balance. Explain that to keep it balanced you must add or subtract the same weight from both sides. Point out that an equation works the same way. The equal sign separates the two sides of the balance. The quantities on each side of the equal sign are equal.

In algebra, a letter is used to represent an unknown quantity. Have students find the formulas for area and volume in Chapter 16. Have them choose letters to represent the words in the formulas and write the equations.

Example: Area = length × width $A = l \times w$.

Words to Know
Review with the students the Words to Know on page 393 of the Student Edition. Have students write the word and its definition on one side of an index card and an example on the other side of the card. Students can use the cards as a reference.

The following words and definitions are covered in this chapter.

equation a mathematical sentence stating that two quantities are equal

variable a letter that stands for a number

number expression a number or numbers together with operation symbols

solution the value of a variable that makes an equation true

simplify to write a shorter or easier form of an expression; or to find its value

parentheses ( ) marks around an operation that should be done first

order of operations the specific order to do the four basic operations when more than one operation is in an equation

Equation Project
Summary: Students find examples of equations in various books and discuss their meanings.

Materials: math books, science books, reference books, paper, pencils

Procedure: Students can complete this project in class with all materials provided. Or, students can search for equations and complete their report outside class.

Assessment: Use the Individual Activity Rubric on page xi of this guide. Fill in the rubric with the additional information below. For this project, students should have:
- found five equations, describing each letter correctly.
- explained one use for each equation.

Learning Objectives
Review with the students the Learning Objectives on page 393 of the Student Edition before starting the chapter. Students can use the list of objectives as a learning guide. Suggest that they write the objectives in a journal or use CM6 *Goals and Self-Check* from the Classroom Resource Binder.

After each lesson, have students write an example of the skill they learned under the appropriate objective. Suggest that students use these notes as a learning guide to help them study for the chapter test.

18.1 What Is an Equation?
Student Edition pages 394–395

Prerequisite Skills
- Mastering basic addition, subtraction, multiplication, and division facts

Lesson Objective
- Identify solutions to equations.

Words to Know
equation, variable, solution

Cooperative Group Activity

Solve It

Materials: clear plastic bags, index cards, counters, stapler or tape

Procedure: Organize students into groups of three or four. Give each group at least three plastic bags and three index cards, 20 counters, and a stapler or tape.

- One student in the group writes a basic fact; for example, $3 \times 4 = 12$. Another student replaces one of the numbers with a variable, then writes the equation on an index card: for example, $3 \times a = 12$. The student then passes this card to another group.
- The next group guesses the solution and staples a plastic bag over the variable on the index card. A student places counters in the bag to represent the solution.

- The students who wrote the equation check whether the suggested number is a solution to the equation. They write *Yes* or *No* on the index card.

Customizing the Activity for Individual Needs

ESL To help students understand the word *equation*, ask them to think of another word that has some of the same letters (*equal*). Draw an equal sign. Then write $5 = 4 + 1$ to illustrate that both sides are equal and form an equation.

Learning Styles Students can:

 use Visual 21 *Balance* to represent equations.

 use counters to see that both sides of the equation are the same.

 repeat an equation you first say with the variable, then with the variable replaced by its value.

Reteaching Activity

Place some marbles in a lunch bag and write *x* on the bag. Place this bag along with 5 marbles on one side of a balance scale. Show students how to use the scale to solve for *x*. Place 5 marbles on the other side of the scale. Then have students add marbles one by one until the scale balances. The number of marbles that were added should equal the number of marbles in the bag. Repeat the activity using different numbers of marbles.

Alternative Assessment

Students make up equations with a given solution.

Example: Write two equations that have 4 for a solution.

Sample answers: $x + 5 = 9$; $8 \div x = 2$; $x - 1 = 3$; $3x = 12$

18.2 Using Parentheses
Student Edition page 396

Prerequisite Skills
- Mastering basic addition, subtraction, multiplication, and division facts

Lesson Objective
- Simplify expressions with parentheses.

Words to Know
number expression, simplify, parentheses

Cooperative Group Activity

Place the Parentheses

Materials: index cards, pencils, paper

Procedure: Organize the class into pairs. Give each pair 16 index cards.

- Partners write the numbers 1–10, as well as two each of the following signs: +, −, ×, and ÷ on separate index cards. (If students have difficulty with fractions and decimals, omit the sign.)
- Partners shuffle their number cards and choose three cards. They shuffle the operation cards and choose two. Students line up the five cards so they alternate: number, sign, number, sign, number.
- Partners write their number expression twice. In the first expression, they put parentheses around the first two numbers. In the second expression, they put parentheses around the last two numbers; for example: $(3 + 2) \times 7$ and $3 + (2 \times 7)$. Then students simplify each expression and record the results.
- Students repeat the activity several times. Volunteers explain how parentheses can affect the value of a number expression.

Customizing the Activity for Individual Needs

ESL Pair students to review the words *number expression* and *parentheses*. Have them identify these words in the exercises on page 396 of the Student Edition. Discuss the difference between an *expression* and an *equation*, and between *solve* and *simplify*.

Learning Styles Students can:

 write each step in the simplification process.

 write number expressions on the chalkboard and simplify.

 say number expressions aloud and simplify.

Reinforcement Activity

Have students work in pairs. Each student writes a number expression with three terms and two operations. They put parentheses around two of the terms and simplify. Students then rewrite the expression *without* parentheses and write the simplified answer next to it. Have the students exchange papers. Students try to reinsert the parentheses so the expression equals the simplified answer.

Alternative Assessment

Students can identify which operation is to be completed first in an expression with parentheses.

Example: How can you simplify 3 (8 − 5)?

Answer: Subtract 5 from 8; multiply the difference by 3.

ON-THE-JOB MATH
Computer Programmer
Student Edition page 397

Lesson Objectives
- Apply algebra to computer programming.
- Communication Skill: Explain the steps in a computer program.
- Workplace Skill: Understand a computer program written in Basic.
- Workplace Skill: Write a computer program in Basic.

Activities

Act It Out

Materials: none

Procedure: Here is another example of a Basic computer program that changes one number into another number.

```
10 PRINT "Type a number from 1 to 10" for x
20 LET a = x * x
30 LET b = 100 - a
40 LET c = b + 10
50 PRINT "We changed your number to:"
60 PRINT c
```

- Have students choose a number for x.
- Then have them follow the steps in the program above to find the number that will be printed at the end of the program. For example, if x = 5, the number printed will be 85.

Tools of the Trade

Materials: paper, pencils

Procedure: Organize the class into small groups.
- Groups study the computer program on page 397 of the Student Edition and in the Act It Out activity above.
- Students write their own computer program that changes one number into another number.
- Groups exchange programs to check and solve.

Write a Business Letter

Materials: telephone books, paper and pen or computer

Procedure: Have students write letters to various appliance manufacturers. Ask them to find out:
- what products incorporate an electronic computer or computer chip.
- how the computer chip helps run the appliance.
- how the chip is programmed to do its job.

Practice

Have students complete the On-The-Job Math Exercise 233 *Computer Programmer* from the Classroom Resource Binder.

Internet Connection

The following Web sites will give students more information about the job of a computer programmer:

Bureau of Labor Statistics, Occupational Outlook Handbook
    http://stats.bls.gov/oco/ocos110.htm
International Programmer's Guild
    http://www.ipgnet.com

18.3 Order of Operations
Student Edition pages 398–399

Prerequisite Skills
- Mastering basic addition, subtraction, multiplication, and division facts

Lesson Objectives
- Simplify expressions using the order of operations.
- Life Skill: Use the order of operations to find the cost of multiple items.

Words to Know
order of operations

Cooperative Group Activity

Who Goes First?

Materials: playing cards, pencils, paper

Procedure: Have students work in pairs. Give each pair a deck of playing cards. Write on the chalkboard: Jack = + ; Queen = − ; King = × ; Ace = ÷
- Students separate aces and face cards in one pile and the numbers 2–10 in another pile.

- Students take turns drawing four number cards and three *operation* cards. They arrange them to form a number expression. They write the expression on a sheet of paper.
- One partner does multiplication and division. The other partner does addition and subtraction. Students decide who must go first, then simplify the expression.
- Pairs exchange roles and repeat the activity.

Customizing the Activity for Individual Needs

ESL Review the operations and their symbols to help students understand *order of operations*. Have students read example problems aloud, pointing out the operation symbols and circling the operation that is to be done first.

Learning Styles Students can:

 use a colored pencil to circle the part of the expression that is to be simplified first.

 keep the rules for the order of operations on an index card available for quick reference.

 learn a jingle that tells the order of operations, such as *my dear aunt Sally* for *multiply, divide, add, subtract*.

Enrichment Activity

Have students try to write expressions equal to each of the numbers 1 to 10, using three different one-digit numbers and two different operations.

Example: $1 = 1 - 1 + 1$.

Alternative Assessment

Students can tell what operation is to be completed first to simplify an expression.

Example: To simplify $14 - 2 \times 5$, what operation do you perform first?

Answer: First multiply 2 times 5.

18.4 Solving Equations with Addition and Subtraction
Student Edition pages 400–401

Prerequisite Skills
- Adding and subtracting whole numbers

Lesson Objectives
- Solve equations with addition and subtraction.
- Life Skill: Use algebra to solve age problems.

Cooperative Group Activity

Number Mysteries

Materials: small self-stick notes, paper, pencils

Procedure: Have students work in pairs. Give each pair four self-stick notes.

- Students work together to write a number sentence using addition or subtraction: for example, $24 - 13 = 11$.
- They write their number sentence on a piece of paper with the numbers large enough for a self-stick note to cover just one number. Students then place a self-stick note over the *first* number in the number sentence and write a variable on the self-stick note. This is a *number mystery*.
- Students exchange their *number mystery* with another pair and solve the equation they are given. Students check their work by lifting the self-stick notes.
- Students repeat the activity.

Customizing the Activity for Individual Needs

ESL To help students understand the meaning of *undo*, have them act out some doing and undoing activities such as, putting on a hat and taking off the hat.

Learning Styles Students can:

 write the addition or subtraction needed to undo the given operation in a different color.

 model problems with counters and manipulate counters as needed.

 describe to a partner the process for solving addition and subtraction equations. The partner repeats it to see if it works.

Enrichment Activity

Provide students with paper clips of two different colors. Tell them that one color represents a value for x and the other a value for y. Have students find all the different ways they can find solutions for $x + y = 9$.

Alternative Assessment

Students can explain what operation to use to solve an equation.

Example: What operation do you undo to solve the equation $y - 4 = 11$? How?

Answer: Undo subtracting 4 by adding 4.

18.5 Solving Equations with Multiplication and Division
Student Edition pages 402–403

Prerequisite Skills
- Multiplying and dividing whole numbers

Lesson Objectives
- Solve equations with multiplication and division.
- Life Skill: Use algebra to solve age problems.

Cooperative Group Activity

More Mysterious Numbers
Materials: small self-stick notes, paper, pencils

Procedure: Have students work in pairs. Give each pair four self-stick notes.
- Students work together to write a number sentence using multiplication or division: for example, $27 \div 3 = 9$.
- They write their number sentence on a sheet of paper with the numbers large enough for a self-stick note to cover just one number. Students then place a self-stick note over the *first* number in the number sentence. They write a variable on the self-stick note. This is a *number mystery*.
- Students exchange their *number mystery* with another pair and solve the equation they are given. Students check their work by lifting the self-stick notes.

Customizing the Activity for Individual Needs
ESL Have pairs of students take turns verbalizing the steps used in solving a multiplication or division equation.

Learning Styles Students can:

 write the multiplication or division needed to undo the given operation in a different color.

 model problems with counters and manipulate counters as needed.

 record the process for solving multiplication and division equations and listen to it.

Enrichment Activity
Have students fill in the blanks in the following sentences as you read them aloud. Then encourage students to write and solve an equation for each problem.

"I am _____ years old. How many years will it be before I turn _____ years old?"

"I am twice as old as _____. I am _____ years old. How old is _____?"

Have students make up their own age problem to solve. Suggest they look at the Everyday Problem Solving features on pages 401 and 403 of the Student Edition for ideas.

Alternative Assessment
Students can explain what operation to use to solve an equation.

Example: What operation do you undo to solve the equation $y \div 4 = 3$? How?

Answer: Undo dividing by 4 by multiplying by 4.

18.6 Problem Solving: Using a One-Step Equation
Student Edition pages 404–405

Prerequisite Skills
- Multiplying and dividing whole numbers

Lesson Objectives
- Use a one-step equation to solve word problems about distance, rate, and time.
- Life Skill: Make a table to solve distance problems.

Cooperative Group Activity

Pick-a-Problem
Materials: blue, white, and pink index cards; paper; pencils

Procedure: Organize students into groups of three. Give each group four of each color index card. Write the distance formula in words and in symbols on the chalkboard.

distance = rate × time $d = r \times t$

- One student in each group writes four different *rates* on separate blue index cards; for example, 55 mph.
- Another student writes four different *times* on separate white index cards: for example, 2 hours.
- The third student writes four different *distances* on separate pink index cards: for example, 250 miles.
- Students shuffle all 12 cards and choose two cards of different colors. They replace the variables in the formula with the chosen values. Students may use a calculator to solve for the third value.

They round decimals to the nearest tenth, and record their work on paper.

- Students continue to choose two cards and solve for the third value until all the cards are used.

Customizing the Activity for Individual Needs

ESL The word *formula* can be found outside mathematics, such as a formula in science or baby milk. Discuss the different situations in which the word can be used. Point out that in each case a *formula* is an established set of rules or process.

Learning Styles Students can:

 use a graphic representation to help fill in the values, for example: $\underline{\text{distance}} = \underline{\text{rate}} \times \underline{\text{time}}$
$d = r \times t$

 draw a representation as above on the chalkboard, then tape index cards in their appropriate blanks.

 tell which parts of the distance formula are given in the problem and which part needs to be found.

Reinforcement Activity

Students can make a table and use guess and check to solve for an unknown in an equation. Have students make a table to solve the following distance problem:

Becky walked a distance of 12 miles. She walked at a rate of 4 miles per hour. How long did it take her to walk this distance?

Time (t)	4 mph $\times t$	Distance
1	4×1	4
2	4×2	8
3	4×3	12

Alternative Assessment

Students can write the equation needed to solve a distance problem.

Example: What equation would you use to solve the following problem? What are you solving for?

A car traveling at 55 mph goes 165 miles. How long does the trip take?

Answers: $165 = 55t$; solve for t.

18.7 Solving Equations with More than One Operation
Student Edition pages 406–408

Prerequisite Skills

- Adding, subtracting, multiplying, and dividing whole numbers

Lesson Objectives

- Solve equations with more than one operation.
- Life Skill: Use algebra to solve problems involving taxi fares.

Cooperative Group Activity

Equation Families

Materials: index cards, paper, pencils

Procedure: Have students work in groups of three. Give each group 12 index cards. Write the following related equations on the chalkboard.

$5x + 10 = 15$ $\qquad$ $5x + 10 = 20$ $\qquad$ $5x + 10 = 25$

- Each student in a group chooses one of the equations and solves the equation for x. *Answers:* $x = 1$; $x = 2$; $x = 3$. Students can check each other's work.
- Group members then use multiplication and either addition or subtraction to create their own family of equations. They write each equation on one side of an index card and the solution to the equation on the other side of the card.
- Groups exchange their equation families and solve.
- Repeat the activity with equations involving division: for example,

$y \div 2 - 6 = 0$ $\qquad$ $y \div 2 - 6 = 4$ $\qquad$ $y \div 2 - 6 = 6$

Answers: $y = 12$; $y = 20$; $y = 24$

Customizing the Activity for Individual Needs

ESL Point out that to solve equations with more than one operation, you *undo* the order of operations. First you undo addition or subtraction, then you undo multiplication or division. Relate this to a real-life activity. First you put on your socks, then your shoes. To *undo* this, you take off your shoes and then your socks.

Learning Styles Students can:

 check that the value for the variable is correct by substituting the value into the equation.

 use counters to model equations.

read aloud equations, then ask a partner what needs to be undone first.

Enrichment Activity

Have students set their own taxi fares by choosing an initial charge and a charge per mile or part of a mile: for example, $1.25 initial charge and $.80 per mile. Then have them write an equation to represent the cost of a taxi ride: for this example, $c = \$.80m + \$1.25$, where c is the cost and m is the number of miles traveled. Have students solve for c by finding the *cost* of traveling a certain number of miles. Then have students solve for m by finding the number of miles they can travel given the cost of a ride.

Alternative Assessment

Students can solve an equation by identifying which operations need to be undone.

Example: What operations do you undo to solve the equation $3x - 8 = 1$? How?

Answer: Undo the subtraction by adding 8. Then undo the multiplication by dividing by 3. The result is $x = 3$.

USING YOUR CALCULATOR
Checking Solutions to Equations
Student Edition page 409

Lesson Objective
- Life Skill: Use a calculator to check solutions to equations.

Activity
Am I Right?
Materials: calculators with algebraic logic, paper, pencils

Procedure: Have students work in pairs.
- One player writes an equation using a variable and two operations. The player also gives the solution for the equation. The solution he or she gives may either be *correct* or *incorrect*: for example, $2y + 7 = 11$; $y = 2$
- The second player looks at the solution. Using mental math, the player decides whether he or she thinks the solution is correct or incorrect. The player then uses a calculator to check the solution. If the player guessed correctly that the solution was correct or incorrect, he or she scores 1 point.
- Partners reverse roles and repeat the activity.
- The first player to score 10 points wins.

18.8 Problem Solving: Using a Two-Step Equation
Student Edition pages 410–411

Prerequisite Skills
- Adding, subtracting, multiplying, and dividing whole numbers

Lesson Objectives
- Use a two-step equation to solve word problems about renting an item by the day.
- Life Skill: Make a table to solve rental problems.

Cooperative Group Activity

Pick-a-Problem Revisited
Materials: blue, white, pink, and yellow index cards; paper; pencils

Procedure: Write the rental formula in words and in symbols on the chalkboard.

Cost = rate per day × number of days + fee
$$C = r \times d + f$$

Organize students into groups of four. Give each group four of each color index card.
- One student in each group writes four different *costs* on separate blue index cards: for example, *Cost: $175*.
- Another student writes four different daily *rates* on separate white index cards: for example, *$15 per day*.
- A third student writes four different *days* on separate pink index cards: for example, *4 days*.
- The fourth student writes four different *fees* on separate yellow index cards: for example, *Fee: $18*.
- Students shuffle all 16 cards and choose three cards of different colors. They replace the variables in the formula with the chosen values. Then they solve for the missing value. They record their work on paper.
- Students continue to choose three cards and solve for the missing value until all the cards are used.

Customizing the Activity for Individual Needs
ESL To help students understand the words in the formula *Cost = rate × number of days + a fee*, dissect the formula so the students see each part. *Cost* is the *total cost*, not the cost per day. The *rate* is the amount charged daily. The number of days depends on the customer's needs. The *fee* does not

change. Write each part on an index card as it is explained, piecing it together in its entirety when your explanation is completed.

Learning Styles Students can:

 use a graphic representation to help fill in the value: for example,

Cost = <u>rate per day</u> × <u>number of days</u> + <u>fee</u>
c   =   r   ×   d   + f

 draw a representation as above on the chalkboard, then tape index cards in their appropriate blanks.

 explain which parts of the formula are given in the problem and which part needs to be found.

Reteaching Activity

Use colored chalk to color-code each part of the formula. Provide students with the same color highlighters. Have students highlight the numbers and the key words (day, cost, fee, rate) on a copy of the practice problems before solving them.

Alternative Assessment

Students can solve word problems by choosing the correct equation from a list.

Example: Which equation would you use to solve the following problem? What is the solution?

The cost to rent a house is $50 per day plus a renter's fee of $100. Kathy paid $250 to rent a house. For how many days did she rent the house?

  $250 = $100 × d + $50
  $250 = $50 × d + $100
  d = $50 × $100 + $250

Answers: $250 = $50 × d + $100; 3 days

Closing the Chapter
Student Edition pages 412–413

Chapter Vocabulary

Review with the students the Words to Know on page 393 of the Student Edition. Then have students quiz each other in pairs.

Have students copy and complete the Vocabulary Review questions on page 412 of the Student Edition.

For more vocabulary practice, have them complete Vocabulary 225 *Algebra* from the Classroom Resource Binder.

Test Tips

Have pairs of students take turns reading one of the test tips and showing how to solve a problem by using that tip.

Learning Objectives

Have students review CM6 *Goals and Self-Check* from the Classroom Resource Binder. They can check off the goal they have reached. Note that each section of the quiz corresponds to a Learning Objective.

Group Activity

Summary: Students investigate the cost of renting an item for different periods of time and write a summary of their findings.

Materials: Yellow Pages of a local calling area, paper, pencils

Procedure: Students can inquire about rental costs for car rentals or equipment.

Assessment: Use the Group Activity Rubric on page xii of the Classroom Resource Binder. Fill in the rubric with the additional information below. For this activity, students should have:

- recorded rental fees and calculated the total cost for the three time periods.
- decided the least expensive way to rent the item and written a paragraph.

RELATED MATERIALS See the unit overview page for other Globe Fearon books that can be used to enrich and extend the material in this unit.

Assessing the Chapter

Traditional Assessment

Chapter Quiz
The Chapter Quiz on pages 412–413 of the Student Edition can be used as either an open- or closed-book test, or as homework. The quiz can be used to identify concepts in the chapter that students need to review and practice.

Chapter Tests
Use Chapter Test A Exercise 235 and Chapter Test B Exercise 236 *Algebra* from the Classroom Resource Binder to further assess mastery of chapter concepts.

Additional Resources
Use the Resource Planner on page 218 of this guide to assign additional exercises from the Classroom Resource Binder and Workbook.

Alternative Assessment

Interview
Write a two-step equation on a piece of paper. Ask:
- Which operation should you undo first?
- How do you undo that operation?
- Which operation should I undo second?
- How do I undo that operation?
- How could I check to see if the answer is the solution to the equation?

Application
Have students role-play that they are renting an item of their choice. Have them make a sign showing the fee and the daily cost. Then have them write two word problems that can be answered using the information on the sign. After solving their problems, post the sign with the problems on a bulletin board for other students to solve.

Unit Assessment

This is the last chapter in Unit 5, Algebra. To assess cumulative knowledge and provide standardized-test practice, administer the practice test on page 414 of the Student Edition and the Unit 5 Cumulative Test, pages T15, T16 from the Classroom Resource Binder. These tests are in multiple-choice format. A scantron sheet is provided on page T2 of the Classroom Resource Binder.

Index

A
Algebra, 217–227
   integers, 207–216
   order of operations, 221–222
   solving equations, 222–225
Average, 164–165

C
Customary Measurement, 170–179
   temperature, 177
   time, 175–177
   units of, 172–174,

D
Decimals, 115–130
   adding, 121
   changing to fractions, 127–128
   changing to percents, 139
   dividing, 124–126
   multiplying, 122–124
   reading and writing, 119
   subtracting, 121–122

F
Fractions, 71–113
   adding, 103–104, 107
   changing to decimals, 128
   changing to percents, 138–139
   dividing, 93–94
   finding common denominators, 78, 79
   lowest terms, 76–77
   multiplying, 89–92
   subtracting, 103, 104, 108–110

G
Geometry, 190–205
   angles, 193–196
   area
      of circles, 201–202
      of parallelograms, 199
      of squares and rectangles, 198
      of triangles, 200
   circumference, 201
   perimeter, 197
   volume
      of prisms, 203
      of cylinders, 203–204
Graphs, 158–166
   bar, 159–160, 162
   circle, 163–164
   histograms, 165–166
   line, 160–162
   pictographs, 158
Greatest common factor (GCF), 64

I
Integers, 207–216
   adding, 210–211, 215
   dividing, 214
   multiplying, 213
   subtracting, 211–212, 215

L
Least Common Multiple (LCM), 64, 65

M
Metric Measurement, 180–189
   comparing to customary, 186–187
   units of, 182–186, 187–188
Mixed numbers
   adding, 104, 108, 110
   changing to fractions, 81, 82
   dividing, 96–98
   multiplying, 92–93
   subtracting, 105–106, 110–111

P
Percents, 131–44
   changing to decimals, 134v135
   commissions, 137
   discounts, 136–137, 142
   finding the part, 135, 143
   finding the percent, 140, 143
   finding the whole, 141–142, 143
   percent increase or decrease, 140–141
   sales tax, 135–136
Prime numbers, 65–66
Probability, 166–168
Proportions, 148–149, 152

W
Whole numbers, 1–70
   adding, 13–22
   comparing and ordering, 8–9
   dividing, 46–58
   multiplying, 34–45
   place value, 6–8
   rounding, 10–11
   subtracting, 23–33
Word Problems
   clue words, 18–19, 28, 41–42, 52–53
   choose the operation, 32, 54–55
   extra information, 68–69
   multi-part problems, 111–112, 127
   reasonable answer, 98
   patterns, 83–84
   solve a simpler problem, 95–96
   two-part problems, 44–45, 187–188
   using an equation, 223–224, 225